Host-Pathogen Interactions in Plant Disease

Host–Pathogen Interactions in Plant Disease

J. E. VANDERPLANK

Department of Agricultural Technical Services
Plant Protection Research Institute
Pretoria, South Africa

1982

ACADEMIC PRESS

A Subsidiary of Harcourt Brace Jovanovich, Publishers

New York London
Paris San Diego San Francisco São Paulo Sydney Tokyo Toronto

ACADEMIC PRESS, INC.
111 Fifth Avenue, New York, New York 10003

United Kingdom Edition published by
ACADEMIC PRESS, INC. (LONDON) LTD.
24/28 Oval Road, London NW1 7DX

Library of Congress Cataloging in Publication Data

Vanderplank, J. E.
Host-pathogen interactions in plant disease.

Bibliography: p.
Includes index.
1. Plant diseases. 2. Host-parasite relationships.
3. Host plants. 4. Micro-organisms, Phytopathogenic.
I. Title.
SB731.V264 632'.3 81-15063
ISBN 0-12-711420-3 AACR2

PRINTED IN THE UNITED STATES OF AMERICA

82 83 84 85 9 8 7 6 5 4 3 2 1

Contents

6 The Gene-for-Gene Hypothesis

7 Some Thermodynamic Background

8 Continuously Variable Resistance to Disease

11 The Spread of Disease

Preface

This work analyzes and updates a wealth of information that has not previously been recorded in other books or reviews. Some of it comes from down-to-earth surveys of disease in the field. The analysis of these surveys not only explains details of host–pathogen interactions that were hitherto obscure, but also indicates the direction for future research. Other data, from original papers, have now been coordinated for the first time and organized in such a way as to suggest new areas for research. There are more than fifty new tables, integrating data from rich and varied sources, relating them to the general principles of host–pathogen interaction.

Host–pathogen interaction underlies all infectious disease, thus this work is primarily for plant pathologists and plant breeders concerned with the control of disease. How to manipulate the host and, indirectly, the pathogen so as to control disease is discussed. It has been found that virulence in fungi has a consistent structure of its own, and this finding opens a whole new area of mycology which relates to plant pathology. Agriculturally it is perhaps the most useful finding in general mycology in decades. To the plant pathologist's armory of chemicals, crop rotation, sanitation, isolation, quarantine, clean seed, controlled environment, and resistant cultivars it adds a new weapon: the consciously ordered restriction of virulence. Other chapters analyze, biometrically and genetically, records of disease resistance that time has shown to be stable in an effort to determine what makes the resistance stable. For biochemists the core of host–pathogen interaction is how variation is stored chemically and how variation in host and pathogen is mutually recognized. Evidence that this variation is often enormous accumulates yearly, and must increasingly influence theories on biochemical plant pathology. For epidemiologists much is discussed that is new. There is an analysis of the structure of epidemics based on three fundamental variables: the initial inoculum, the progeny/parent ratio, and the latent period. The spotlight falls on the progeny/parent ratio, which allows a new synthetic treatment of epidemiology.

The chapters are essentially self-contained and designed to follow one another in organized sequence. The book is at least four-fifths new in that most of the topics have not appeared in my previous books nor been discussed by others. Repetition has been reduced to the minimum needed for continuity and clarity. Where information from previous books is relevant, simple reference to the earlier work has been deemed adequate.

J. E. Vanderplank

1

Introduction

The past decade has been one of great advances in the science of plant pathology, and some of them have been unusually exciting. Viroids were discovered, and with their discovery the range of plant pathogens has been extended to its lowest conceivable limit: pathogens that are single chemical molecules able to get themselves reproduced by the host plant. The crown gall bacterium was analyzed genetically to reveal its guile. It transfers some of its DNA to the host's genome, which causes the host plant to produce atypical amino acids that the bacterium can assimilate but the host plant cannot. Exciting advances like these along narrow fronts have been accompanied by the steady accumulation of knowledge over a broad front, and plant pathology in almost all its various divisions from bacterial genetics to curative fungicides is today a much better science than it was ten years ago. This is particularly true of matters concerning host/pathogen relations, which are the topic of this book.

Chapter 2 starts by examining the pathogen; its topic is the structure of virulence. This is a topic about which little could have been written ten years ago. The pathogen adapts, or tends to adapt, itself to the host. This much was well known, and adaptation is indeed the basis of all disease. It was also known that there are constraints on adaptation which go under the general name of homeostasis or stabilizing selection. But is was not known that virulence has its own structure only indirectly influenced by the host. If on the gene-for-gene hypothesis we identify virulence in the pathogen by the corresponding resistance gene in the host, virulence for some resistance genes strongly dissociates and for other resistance genes strongly associates, in a way that seems to have little to do with pathogen/host adaptation. The evidence for this is both massive and consistent, and it is possible to classify resistance genes in the host according to how the corresponding virulences in the pathogen behave. In this behavior we have a useful weapon for the control of disease by grouping in the host the best combination of resistance genes. This weapon has often been used unconsciously in the past, but now with more exact planning it is possible to reduce waste of resistance genes and of the plant breeder's time.

Chapter 2 leads to Chapter 3 which takes up the matter of races of the pathogen defined by their virulence or avirulence for particular resistance genes in the host. If there is a gene for resistance in the host plant, then, ignoring the possibility of intermediates, there could potentially be two races of the pathogen, one virulent and the other avirulent for this gene. If there were two resistance genes, the number of potential races would be four: if there were three resistance genes, the number of potential races would be eight; and so on, with the number of potential races always increasing geometrically. This was first brought out clearly in a proposal by Black *et al*. (1953) for an international system of classifying races of *Phytophthora infestans*. At that time there were four known *R* genes in the potato, and Black *et al* (1953) deduced the existence of 16 races of *P. infestans*. These have all been found, and the system of classification is now taken for granted. With the further discovery of genes *R5, R6,* etc., the number of races had to be extended, geometrically, from 16 to 32, from 32 to 64, etc. This was accepted without special comment: there are only eleven *R* genes identified in *Solanum,* and the corresponding 2048 races of *P. infestans* are still within conceivable limits. What now boggles the mind are the numbers, literally inconceivable, involved in some other host/pathogen systems. There are now at least 30 known *Sr* genes in wheat for resistance to stem rust, so that with all possible permutations and combinations of virulence and avirulence there are (in approximate numbers) potentially a billion races of *Puccinia graminis tritici,* even if one ignores intermediate interactions. This figure of a billion races is itself hard enough to swallow, but the extension is still harder. If there were 31 *Sr* genes, the number of races would potentially be two billion. That is, the discovery of the thirty-first gene would raise the number of potential races from one to two billion. One single newly discovered gene locus in wheat would have to be accommodated by the creation of a billion new races of *P. graminis tritici*. Of course, in practice the figure cannot be realized, among other reasons because it is far beyond the range of experimental feasibility. This is, however, beside the point that taxonomically we have arrived at a situation in which a single gene locus can potentially create a billion new taxa, a situation foreign to all established taxonomy. Fortunately we can now turn to computers, and rely on their memory to extract information on those particular virulence/avirulence combinations that are the purpose of the survey. At the same time advantage could be taken of computerization to determine environmental effects on virulence/ avirulence association, because what clearly emerges from the evidence is that these effects are strong.

Chapter 4 takes up the host plant's effect on virulence/avirulence associations in the pathogen. The boom-and-bust cycle enters the discussion here. The plant breeder introduces resistance in the host to disease, and starts the boom; the pathogen adapts itself to the resistance, and starts the bust. Until recently the emphasis in the literature was on the source of the variation the pathogen needs for its adaptation. Variation was traced to sexual recombination, as in *Puccinia graminis* on barberry bushes, and to a variety of parasexual and asexual processes that can change avirulence to virulence. The modern trend, as shown in the review by Person *et al.* (1976), is to accept that mutation from avirulence to virulence in the pathogen is frequent enough to supply the necessary variation, and to put the emphasis on whether the virulent mutants will survive. Both mutation to virulence and the survival of virulent mutants are proper topics for study. Recent research in Australia and elsewhere suggests that mutation rates to virulence are highly variable; the pathogen mutates from avirulence to virulence with ease for some resistance genes and with difficulty for others. To understand the survival of virulent mutants one must distinguish between genetic mutation, phenotypic mutation, and epidemiological mutation. Genetic mutation from avirulence to virulence is an appropriate change in the DNA, and concerns the pathogen alone. Phenotypic mutation to virulence in the pathogen, derived of course from genetic mutation, concerns both pathogen and host, in that the phenotype of the pathogen cannot reflect mutation to virulence except in combination with the host. Epidemiological mutation too concerns both pathogen and host, but with both of them in populations as distinct from individual phenotypes. Epidemiological mutation, which is the mutation relevant to boom-and-bust cycles, requires that the pathogen should be fit to survive in the host population, fitness here being taken in its Darwinian sense of success in leaving progeny. The three sorts of mutation occur in decreasing order of frequency, genetic mutants to virulence not necessarily being observed as virulent phenotypic mutants, and virulent phenotypic mutants not necessarily surviving and increasing in the host population to cause epidemics.

Chapter 5 discusses horizontal resistance and vertical resistance as biometric concepts. In a two-variable system, with host and pathogen the only variables, there can be two and only two forms of resistance in the host, horizontal resistance being defined by the main effect and vertical resistance by interaction, i.e., by the differential effect. The term, main effect, is used strictly in its biometric sense. A confusion in the literature seems to have arisen from the unwarranted assumption that main effects and interactions cannot coexist and that there is consequently an either/or choice of source of resistance. In reality, horizontal and vertical resistance can occur mixed in any proportion, just as in the analysis of variance it is a basic concept that main effects and interactions coexist or can coexist in any proportion. Discussing resistance against a background of the analysis of variance highlights the obvious point that horizontal and vertical resistance cannot be distinguished unless there are enough degrees of freedom to detect interaction.

Nearly 40 years ago Flor proposed his gene-for-gene hypothesis: for every gene for resistance in the host there is a corresponding and specific gene for virulence in the pathogen. The hypothesis is the topic of Chapter 6, and the literature about it is large. Modern findings suggest that some minor amendments may be needed, but they do not affect the general validity of the hypothesis. The hypothesis remains the cornerstone of genetical plant pathology. In diseases to which it is applicable there is at least some light on host/pathogen specificity. In diseases to which it is inapplicable, with the exception of diseases caused by specific toxins, there are few real clues to explain the host ranges of pathogens. What concerns us most in Chapter 6 is the accumulated evidence about the sheer weight of numbers of genes now known to be involved. Reference has already been made to 30 *Sr* genes in wheat for resistance to stem rust. (There are almost certainly many more.) To this number one must add the numbers of genes in wheat for resistance to leaf rust, stripe rust, powdery mildew, loose smut, the bunts, and the Hessian fly, all thought to be on a gene-for-gene basis. The total, probably much more than 60, raises the question how host and pathogen store this variation in such a way that each can recognize the other's. The storage itself poses no difficulty in modern genetics: it is in the DNA (or RNA of some viruses). But, at least with eukaryotic pathogens, recognition is not at DNA level; and the capability both to vary adequately and to have this variation mutually recognized is likely to be found only in RNA, proteins, and saccharides. We have here a detective story, with five clues. Susceptibility is specific, and in gene-for-gene systems requires both a specific inducer and a specific receptor. Gene-for-gene parasitism is biotrophic, at least for some time after infection. Large numbers of genes may be involved in both host and pathogen, implying massive variation precisely recognized. When different recognition systems are involved, there are only pseudoalleles, not alleles. Susceptibility in gene-for-gene disease is endothermic. Evidence for these clues is given in detail in Chapters 6 and 7; together, the clues seem to lead in only one direction.

Resistance to disease in a gene-for-gene system is qualitative and varies discontinuously. Equally important is quantitative resistance that varies continuously; this is the topic of Chapter 8. In the literature most of this continuously variable resistance has been called polygenic, and has been regarded as relatively stable and difficult for the pathogen to overcome. In reality, little or none of this resistance is polygenic, and the notion of polygenic resistance has arisen from a confused argument. True polygenic resistance, i.e., resistance conditioned by many genes each of small effect, must necessarily be continuously variable. It is the converse, that continuously variable resistance is necessarily polygenic, that is false. Most of what we have been calling polygenic resistance is for practical purposes oligogenic; and the difference between what we have been calling oligogenic resistance and what we have been calling polygenic resistance is mainly the difference between dominance variance and additive variance. The literature leaves little doubt about this. A change of emphasis, from a mistaken concept of polygenic resistance and an unfounded belief about safety in gene numbers, makes it easier to plan and carry out programs for the breeding of plants with resistance to disease. It is also likely to facilitate biochemical research into the nature of quantitative resistance.

Chapter 9 centers primarily around exceptions to the general rule that vertical resistance delays the start of an epidemic and horizontal resistance slows the epidemic down. Ontological and seasonal changes can cause horizontal resistance to be manifested as a postponement of adult-plant susceptibility, and this leads secondarily to the study of ontological and seasonal effects on resistance and susceptibility of the host.

Mathematical models of epidemics of plant disease originated mainly around a few diseases, such as the cereal rusts, potato blight, and some virus diseases, about which there were adequate data for discussion. It happens to be a feature of these diseases that, given unlimited time to develop, they go to completion, i.e. disease, given time, ends at or very near 100%. This feature was built into the discussions, e.g., in the use of logits to provide a simple measure of the infection rate. There are however other diseases that seldom if ever proceed to 100% infection; they reach a peak often far short of 100% early in the season, after which the percentage infection begins to decline. Powdery mildew of wheat and barley often behaves in this way; Chapter 10 is mainly about diseases like it. The same model stated as a differential equation is used to cover all disease; the variables in the equation determine whether there is a peak or not. Three variables, the initial inoculum, the progeny/parent ratio, and the latent period, determine the form of the disease progress curve; and the effect of the progeny/parent ratio and the latent period is discussed in some detail.

The increase or decrease of disease in time and space summarizes the collective effect of every host/pathogen interaction. It is also one of the most difficult topics in plant pathology, especially in relations to change in space. Historically, what seemed to be the obvious starting point for research on spatial change was the dispersal of pathogens; and few topics in plant pathology have over the years received more detailed and expert attention than pathogen dispersal, be it the dispersal of windblown or rain-splashed fungus spores or bacterial cells, vector-borne viruses, or swimming nematodes. In point of fact, research from this starting point has run into difficulties when applied to polycyclic disease. In Chapter 11 we start differently, using the one fact constant for all plant disease: sandwiched in between an area of polycyclic disease and the area of healthy plants into which disease is spreading, is inevitably an area of monocyclic disease. From this it can be deduced that gradients along which disease spreads depend not only on the pattern of dispersal of pathogens and on the amount of disease, but also on the speed with which this amount is reached, i.e., on the infection rate itself.

Plant pathology is a practical science, and in the long run the amount of financial support and recognition it gets will depend on how useful it is. In this book those host/pathogen interactions have been chosen for study which seem most directly related to the control of plant disease. Topics specially stressed include ways of curbing the virulence of pathogens and of increasing the resistance of host plants, and disease has been discussed primarily in its epidemiological aspects. Freshness too has been an aim in writing the chapters. With few exceptions the principles discussed have been developed within the past decade, even though some of the experimental data and observations behind them go back for many years and have now been put in a new light.

In the various chapters there is an inevitable concentration on relatively few pathogens, because the experimental evidence discussed in the literature is concentrated around relatively few pathogens. *Puccinia graminis,* for example, is the subject of Chapter 2, because no other pathogen has been surveyed over such a wide area, in such detail, and with such highly developed techniques. There are advantages in the concentration. It knits chapters together, and provides a continuous background. It follows the general trend in science. Geneticists, for example, have given more than proportionate attention to man, fruit flies, maize, bread mold, *Escherichia coli* K12, and phages of *E. coli.* The path to principle is the path away from diffuse narrative about many organisms to concentrated attention on relatively few organisms here and there; and in this book I have as much as possible followed a path of concentration.

2

Virulence Structure of *Puccinia graminis* Populations

2.1 INTRODUCTION

This chapter deals with the structure of virulence in a fungus population. The ability of pathogens to adapt themselves to host plants is immense. When a new cultivar of the host is introduced with vertical resistance, the pathogen all too often matches the resistance with newly accumulated virulence, and the cultivar becomes susceptible. This is a topic that pervades the literature of breeding plants for resistance to disease; it is the topic of Chapter 4. But the ability of pathogens to adapt themselves is not unlimited. There are constraints on adaptation, among which are virulence/avirulence structures in the pathogen population that cannot be distorted without loss of fitness.

The virulence structure of the wheat stem rust fungus *Puccinia graminis tritici* has been revealed in far more detail than that of any other pathogen. Fine work by wheat geneticists, notably in the United States, Canada, and Australia, has made available a series of near-isogenic wheat lines differing from one another in a single gene for resistance. Isolates of *P. graminis tritici* from infected wheat fields can be classified according to which of these lines they can and which they cannot attack; that is, isolates can now be classified according to whether they are virulent or avirulent on plants of particular lines. This development of research is recent, and the sort of surveys of virulence we analyze became possible only in the last decade or so.

Two extensive surveys of *P. graminis tritici* are made annually in North America. One is made on wheat in the United States and Mexico by the United States Department of Agriculture and the University of Minnesota at St. Paul. The other is made in Canada by the Canada Department of Agriculture at Winnipeg. Reference will also be made to the Australian survey conducted by the University of Sydney. The data from these surveys are massive, and are the raw material used to determine virulence structure. In addition evidence on a smaller scale comes from *P. graminis avenae,* the oat stem rust fungus.

Genes in wheat for resistance to *P. graminis* are known as *Sr* genes (*Sr* for stem rust). More than 30 are already known, and they are numbered with international uniformity, thus *Sr5, Sr6,* etc. The corresponding genes in oats are *Pg* genes (*Pg* for *P. graminis*). In this chapter we use no special symbols for virulence in the pathogen. Isolates of the pathogen are recorded simply as virulent or avirulent on plants with a particular *Sr* or *Pg* gene. Alternatively, what comes to the same thing, a particular *Sr* or *Pg* gene is recorded as ineffective or effective against a particular isolate of the pathogen. In the relevant surveys, i.e., those in the United States, Canada, and Australia, results are recorded in a simple binary system, plants being either resistant or susceptible, without (except in very rare circumstances) intermediates being mentioned. For purposes of survey, gone is the complicated system of seven or more categories, from 0 = "immune," through 0;, 1, 2, and 3, to 4 = very susceptible, with one or two plus or minus signs added to get even finer distinctions.

All differences specially referred to in the text are statistically significant, often with *P* values vanishingly small. To indicate the great wealth of evidence numerical data are included either in the body of the tables or in the footnotes.

2.2 THE DISSOCIATION OF VIRULENCE FOR GENES *Sr6* AND *Sr9d* IN CANADA

One of the greater success stories in the history of plant breeding was the development of the wheat cultivar Selkirk by the Canada Department of Agriculture in Winnipeg. Licensed in 1953, it had almost complete resistance to stem rust, which caused it to dominate the spring-wheat fields both south and north of the Canadian frontier. By 1956 it covered 80% of the wheat area of Manitoba, by 1961 86% (Green and Campbell, 1979). Later it started to decline, for reasons unconnected with stem rust, but even as late as 1977 it still covered 10.6% of the area which in absolute terms means hundreds of thousands of hectares.

Consider the background. In 1950 *P. graminis tritici* exploded in the wheat fields of North America. A population of this fungus with a new virulence combination that could attack most of the existing wheat varieties became established in a single season over an area of 3 million square miles (Stakman and Harrar, 1957). For the next three years the epidemic continued unabated. Then the cultivar Selkirk was released. It was resistant, and its resistance to stem rust has continued to the present day. In Canada (but not always elsewhere) its resistance is stable.

Selkirk contains the wheat stem rust resistance genes *Sr6* and *Sr9d,* among others which do not enter our story. In *Puccinia graminis tritici* in Canada virulence on plants with the gene *Sr9d* is common; virulence is far more frequent than avirulence. Virulence on plants with the gene *Sr6* is not uncommon; in most years from 10 to 20% of isolates are virulent on plants with this gene. But combined virulence for the two genes is so rare that it is seldom detected in the annual Canadian surveys. Vanderplank (1975) analyzed earlier data. Other data, for 1972 through 1977, are given in Table 2.1. During these years combined virulence was found only in 1975 when three isolates with combined virulence were collected. From evidence taken as a whole, it seems that from the time that Selkirk was released, only in the two years 1960 and 1975 was combined virulence detected in surveys. This statement is a summary of results with thousands of isolates over the years.

Here we have the explanation of the continuing resistance of Selkirk to stem rust. Adaptation of *P. graminis tritici* to Selkirk would require combined virulence for genes *Sr6* and *Sr9d* apart from virulence for the other resistance genes Selkirk has. But virulence for these two genes is, in Canada, constrained from associating, except rarely and in trace amounts. In the terminology of

TABLE 2.1

The Percentage of Isolates of *Puccinia graminis tritici* in Canada Virulent for Wheat Stem Rust Resistance Genes *Sr6* and *Sr9d,* Singly and in Combination, in the Years 1972–1978

	Year[a]						
Resistance gene(s)	1972	1973	1974	1975	1976	1977	1978
Sr6[b]	16.8	11.3	11.1	14.2	5.3	23.1	17.1
Sr9d[b]	82.8	88.7	92.3	93.6	94.8	72.6	74.1
Sr6 + *Sr9d* expected[c]	13.9	10.0	10.2	13.3	5.0	16.8	12.7
Sr6 + *Sr9d* found[b]	0	0	0	0.9	0	0	0

[a] The total number of isolates was 282, 106, 429, 332, 602, 653, and 304 for the seven years, respectively.

[b] Data of Green (1972b, 1974, 1975, 1976a,b, 1978, 1979) for the years 1972–1978, respectively.

[c] Expected on the assumption that the combination of virulences was selectively neutral and random.

population genetics directional selection toward adaptation of the parasite to the host has failed to overcome the stabilizing selection that has constrained the parasite from change. Stabilizing selection against combined virulence for genes *Sr6* and *Sr9d* has prevented *P. graminis tritici* from colonizing the tens of millions of hectares of Selkirk wheat that have been grown over the years.

The stabilizing selection is explained by the observation of Katsuya and Green (1967) and Green (1971b) that the few isolates of *P. graminis tritici,* with combined virulence for genes *Sr6* and *Sr9d,* were unaggressive and could not maintain themselves in nature. They were unfit, this adjective being used in its Darwinian sense to mean relative incapability for leaving progeny.

2.3 THE ASSOCIATION OF VIRULENCE FOR GENES *Sr6* AND *Sr9d* IN THE UNITED STATES AND MEXICO

It has been fully established that wheat stem rust in Canada starts from spores blown in from the United States. Since the alternative host plants (*Berberis* spp.) have been effectively eradicated in Canada, *P. graminis tritici* cannot survive the Canadian winters. The fungus overwinters in Mexico, Texas, Kansas, and Oklahoma (Rowell and Roelfs, 1971), and through the flight of its uredospores makes its way northward to Canada as fields of wheat grow in the spring and summer. Of necessity, the story of wheat stem rust in Canada starts each year from the south; and our analysis must begin there.

In the United States and Mexico virulence for gene *Sr6* is associated with virulence for gene *Sr9d;* in Canada it is dissociated from it. This is the message of the two surveys; and in these surveys the change at the United States/Canada border is abrupt.

Data are given in Table 2.2. To illustrate them, consider the year 1977. In that year, out of a total of 1207 isolates of *P. graminis tritici* collected in the United States, 24% were virulent for gene *Sr6*. They belonged to three races: QSH, RKQ, and RTQ. These races are also virulent for gene *Sr9d*. That is, virulence for gene *Sr6* in the United States was always associated with virulence for gene *Sr9d*. Consequently the entry in Table 2.2 is 100%. In Canada, in 1977, 23.1% of the isolates of *P. graminis tritici* were virulent for gene *Sr6*. They belonged to races C25, C35, C41, C57, C58, and C63. None of these races is virulent for gene *Sr9d*. The entry in Table 2.2 is therefore 0%. Consider now Canada in 1975. This, it will be remembered, was the only year since 1960 in which associated virulence for genes *Sr6* and *Sr9d* was found in the Canadian surveys. Of the total of 332 isolates of *P. graminis tritici,* 14.2% were virulent for gene *Sr6* and 0.9% (belonging to race C50) were virulent for both *Sr6* and *Sr9d* (see Table 2.1). The percentage of isolates virulent for *Sr6* which were also virulent for *Sr9d* is therefore $0.9 \times 100/14.2 = 6.3$, which is the figure entered in Table

TABLE 2.2

The Percentage of Isolates of *Puccinia graminis tritici* in Mexico, the United States, and Canada Virulent for the Wheat Stem Rust Resistance Gene *Sr6* Which Was Also Virulent for Gene *Sr9d* in the Years 1973–1978[a]

	Year[b]					
	1973	1974	1975	1976	1977	1978
Mexico	100	100	100	100	100	100
United States	100	100	100	100	100	100
Canada	0	0	6.3	0	0	0

[a] Data of Roelfs and McVey (1974, 1975, 1976) and Roelfs *et al.* (1977a, 1978b, 1979a) for the United States and Mexico; data of Green (1974, 1975, 1976a,b, 1978, 1979) for Canada.

[b] The total number of isolates in the samples from Mexico was 58, 81, 143, 138, 32, and 122, and from the United States 1306, 2609, 2418, 1703, 1207, and 790 for the six years, respectively. In the United States the percentage of isolates virulent for gene *Sr6* was 13, 6, 11, 7, 24, and 18 for the six years, respectively. Information for Canada is given in Table 2.1.

2.1. Alternatively, the figure can be gotten directly. Of the 47 isolates virulent for gene *Sr6*, three (i.e., those of race C50) were also virulent for gene *Sr9d*. The percentage of isolates virulent for gene *Sr6* which were also virulent for gene *Sr9d* is therefore $3 \times 100/47 = 6.3$.

We can now consider the data in Table 2.2 more broadly. The Canadian and United States surveys disagree. In Canada virulence for gene *Sr6* is (with a minor exception) dissociated from virulence for gene *Sr9d*. In the United States and Mexico virulence for gene *Sr6* is associated with virulence for gene *Sr9d*. There is no geographical gap in the data. The figures for the United States as a whole apply (so far as our present topic is concerned) also to the North-Central States bordering on Canada. Associated virulence for genes *Sr6* and *Sr9d* is, it seems, shed when *P. graminis tritici* enters Canada. The 49° line that is the border of central United States and Canada divides the immense wheat area of the Great Plains. Wheat cultivars straddle the border. Winds carry uredospores freely without border restrictions. Yet in the surveys *P. graminis tritici* behaves as if it were politically aware. From Mexico to the Canadian border, from the Atlantic to the Pacific coast south of the border, virulence for gene *Sr6* was found during the surveys always in association with virulence for gene *Sr9d*. North of the border it was found only dissociated from virulence for gene *Sr9d*. There is a sharp change northward from association to dissociation. No trace of statistical doubt exists about this; the samples are too large, and the results from year to year too consistent for biometrical error. We must accept the facts of the surveys.

Puccinia graminis tritici sheds its associated virulance for genes *Sr6* and *Sr9d* when it enters Canada. One would have expected the forces of adaptation, that is, directional selection, to have forced *P. graminis tritici* to keep its associated virulence, because this associated virulence is needed if *P. graminis tritici* is to infect Selkirk wheat. The crucial question is not, why did *P. graminis tritici* not develop the virulence needed to attack Selkirk wheat? It is, why, already having associated virulence, does *P. graminis tritici* proceed to shed it just when it is needed? The fact that Selkirk has other resistance genes, *Sr7b, Sr17,* and *Sr23,* does not affect the argument.

Lest it be thought that the sudden reversal at the Canadian border is an artifact resulting from different interpretations of experimental data, let it be said at once that this is not so. The workers responsible for the United States survey have recently tested Canadian isolates along with their own (Roelfs *et al.*, 1979a); and their findings substantiate the data analyzed in this chapter.

2.4 THE ASSOCIATION AND DISSOCIATION OF VIRULENCE FOR GENE *Sr9d,* ON THE ONE HAND, AND GENES *Sr9a, Sr9b,* OR *Sr15,* ON THE OTHER

In relation to virulence for gene *Sr9d,* virulence for genes *Sr9a, Sr9b,* and *Sr15* behaves much like virulence for gene *Sr6,* except that the contrast between association and dissociation is not quite so sharp. Extensive data are given in Tables 2.3, 2.4, 2.5, and 2.6.

Table 2.3, constructed like Table 2.1, shows that in Canada the associated virulence for genes *Sr9d,* on the one hand, and *Sr9a, Sr9b,* or *Sr15,* on the other, is far less than would be expected in the absence of selection pressures, i.e., than would be expected if association and dissociation were random. Tables 2.4, 2.5, and 2.6 tell much the same story about virulence for genes *Sr9a, Sr9b,* and *Sr15* as Table 2.2 told about virulence for gene *Sr6*. In Mexico and the United States virulence for genes *Sr9a, Sr9b,* and *Sr15* exists mostly in association with virulence for gene *Sr9d*. No essential differences were found within the geographical area of Mexico and the United States, but at the Canadian border there is a sharp reversal. Association changes to dissociation; and virulence for genes *Sr9a, Sr9b,* and *Sr15* proceeds northward largely dissociated from virulence for gene *Sr9d*. Some aspects of this story will be taken up later; for the moment we are simply documenting the evidence contained in the literature of the surveys.

No known wheat variety contains combinations of genes *Sr9a, Sr9b,* and *Sr9d*. The association and dissociation of virulence we have been discussing is unlikely to be a direct response to host gene association and dissociation.

TABLE 2.3

The Percentage of Isolates of *Puccinia graminis tritici* in Canada Virulent for Wheat Stem Rust Resistance Genes *Sr9a, Sr9b, Sr9d,* and *Sr15,* Singly and in Combination, in the Years 1972–1977[a]

	Year					
	1972	1973	1974	1975	1976	1977
Sr9d	82.8	88.7	92.3	93.6	94.8	72.6
Sr9a	25.7	21.6	8.7	6.9	5.6	5.2
Sr9b	26.0	21.6	14.8	6.4	6.6	25.2
Sr15	20.6	17.8	17.8	17.5	7.0	28.0
Sr9d + *Sr9a* expected[b]	21.3	19.2	8.0	6.5	5.3	3.8
Sr9d + *Sr9a* found	0	0.9	0.5	1.5	0	0.6
Sr9d + *Sr9b* expected[b]	21.6	19.2	13.7	6.0	6.3	18.3
Sr9d + *Sr9b* found	0	0.9	0.5	1.5	0	0.6
Sr9d + *Sr15* expected[b]	17.1	15.8	16.4	16.4	6.6	20.3
Sr9d + *Sr15* found	2.4	1.0	1.6	1.5	0.2	0.8

[a] Information from the same sources as in Table 2.1.

[b] Expected on the assumption that the combination of virulences was random and selectively neutral.

TABLE 2.4

The Percentage of Isolates of *Puccinia graminis tritici* in Mexico, the United States, and Canada Virulent for Wheat Stem Rust Resistance Gene *Sr9a* Which Was Also Virulent for Gene *Sr9d* in the Years 1973–1977[a]

	Year				
	1973	1974	1975	1976	1977
Mexico	91.0	100	100	100	100
United States	100	100	85.7	100	100
Canada	4.4	5.3	21.8	0	11.8

[a] Data from the same sources as in Tables 2.1 and 2.2. In the United States the percentage of isolates virulent for gene *Sr9a* was 23, 22, 14, 10, and 18 for the years 1973–1977, respectively.

TABLE 2.5

The Percentage of Isolates of *Puccinia graminis tritici* in Mexico, the United States, and Canada Virulent for the Wheat Stem Rust Resistance Gene *Sr9b* Which Was Also Virulent for Gene *Sr9d* in the Years 1973–1977[a]

	Year				
	1973	1974	1975	1976	1977
Mexico	100	100	100	100	100
United States	100	100	100	100	100
Canada	4.4	3.1	23.6	0	2.4

[a] Data from the same sources as in Tables 2.1 and 2.2. In the United States the percentage of isolates virulent for gene *Sr9b* was 13, 4, 12, 7, and 26 for the years 1973–1977, respectively.

TABLE 2.6

The Percentage of Isolates of *Puccinia graminis tritici* in Mexico, the United States, and Canada Virulent for the Wheat Stem Rust Resistance Gene *Sr15* Which Was Also Virulent for Gene *Sr9d* in the Years 1974–1977[a]

	Year			
	1974	1975	1976	1977
Mexico[b]	100	100	100	100
United States[b]	(>90)	100	100	100
Canada	9.2	8.6	4.8	0.5

[a] Data from the same sources as in Tables 2.1 and 2.2.

[b] Estimated approximately on the basis that isolates with the formulas QSH, QFB, QCB, RKQ, RTQ, and RPQ are virulent and those with the formulas TNM, TDM, and TLM are avirulent for gene *Sr15* (Roelfs *et al.*, 1978b).

2.5 THE INVOLVEMENT OF DATES AND TEMPERATURE

Dates and temperature are almost certainly involved in the data. *Puccinia graminis tritici* moves each year from south to north, from the United States to Canada, from winter to summer. South of the Canadian border virulence for gene *Sr6*, when it is present, is associated with virulence for gene *Sr9d*. North of the border it is dissociated from it. But we do not accept that uredospores instantly exchange association for dissociation as they are blown over the border. There is an alternative interpretation. Association is recorded in the United States *at the time of the United States survey;* and dissociation is recorded in Canada *at the time of the Canadian survey*. There is a difference in time as well as in geography between the surveys. The United States surveys record early appearances of stem rust in the annual movement northward; and there are records of early observations of rust (without, however, numerical data). The Canadian survey, being for stem rust at the terminus of its yearly run, is necessarily carried out later.

A time difference is also a temperature difference. At the start of the run, in the south in Texas, the average temperature of the wheat areas during December, January, and February is about 11°C (Wallin, 1964). At the end of the run, in the north, the average temperature is near 20°C; the 20° isotherm for July cuts the Great Plains near the United States–Canada border. Yearly in North America as *P. graminis tritici* moves northward the average temperatures increase by 8° to 9°C.

The range of temperature at which *P. graminis tritici* is found, even in the absence of the alternative host, is much greater than the range at which destructive stem rust epidemics are likely to occur. Winter temperatures in the south are too low for serious stem rust epidemics but they are nevertheless temperatures which *P. graminis tritici* must survive and be genetically adapted to. The United States and Canadian surveys necessarily reflect adaptations to condition at or before the time of the surveys; and differences between surveys may well reflect differences in temperature. The basic assumption here, that adaptation is dynamic and not static, is abundantly clear in comparisons of the two surveys; a static adaptation would need only a single survey.

If our reasoning is correct, populations with relevant virulence associations should be less fit at higher temperatures relative to populations without these virulence associations. By relevant virulence associations, we mean virulence for gene *Sr9d* associated with virulence for genes *Sr6, Sr9a, Sr9b,* or *Sr15*. To anticipate later discussion we include also virulence for gene *Sr9e* associated with virulence for genes *Sr6, Sr9a, Sr9b,* or *Sr15*. Relevant data on the effect of temperature on fitness are meager but, so far as they go, support our contention. Katsuya and Green (1967) investigated the effect of temperature on two races, 56 and 15B-1, of *P. graminis tritici,* using cultures derived from single spores. Susceptible wheat varieties (Little Club, Red Bobs, and Marquis) were inoculated at different temperatures, and the relative abundance of the two races compared after successive generations. Races 56 (= C17) and 15B-1 (= C10) are both avirulent for gene *Sr6,* and both virulent for gene *Sr15*. Virulence for these two genes is therefore outside the comparison. There remain genes *Sr9a, Sr9b, Sr9d,* and *Sr9e*. Race 56 is avirulent for them. Race 15B-1 is virulent for them, and therefore (in terms of our discussion) has virulence associations that ought to reduce relative fitness at higher temperature. This is what is in fact found (see Tables 2.7 and 2.8). At 25°C, sharply, and at 20°C, less sharply, the relative abundance of race 15B-1 in a mixture with race 56 declines with succes-

TABLE 2.7

The Effect of Temperature on the Percentage of Infection by a Culture of Race 15B-1 (Can.) of *Puccinia graminis tritici* Grown for Seven Uredial Generations on Three Susceptible Varieties of Wheat Mixed and in Competition with a Culture of Race 56[a]

	Temperature (°C)		
Generation	15	20	25
0[b]	50	50	50
1	38	33	36
2	44	27	8
3	54	20	17
4	64	10	4
5	61	5	1
6	68	5	2
7	66	4	0

[a] Data of Katsuya and Green (1967). The three susceptible varieties were Little Club, Red Bobs, and Marquis. The data are averages for these three varieties, and were obtained by counting approximately 30,000 infections on the susceptible variety Mindum.

[b] The data are adjusted for an equal (50%) start of the two cultures at generation 0; that is, 50% of the infections were by race 15B-1 and 50% by race 56. The percentage infection by race 56 can be obtained by subtraction from 100; for example, after seven generations at 15°C, 66% of the infections were by race 15B-1, whereas 34% were by race 56; at 20°C, 4% were by race 15B-1 and 96% by race 56, etc.

sive generations. Higher temperatures, 20° and 25°C, favor race 56. Low temperature, 15°C, is neutral or inconsistent in its effect; Table 2.7 shows the proportion of race 15B-1 to increase, and Table 2.8 shows it to decrease, with successive generations at 15°. For a comment on artifacts, see Section 4.6.

These results are supported by the field observations of Roane *et al.* (1960). These workers selected fields of susceptible varieties of wheat in Virginia that had become infected by *P. graminis tritici* from *Berberis* bushes growing near them. The first records were made early in the season, just after infection was first seen on the young plants. The last records were made late, when the fields were nearly ripe. The increase of rust between the early and late collections resulted primarily from an increase within the fields themselves, i.e., during the interval further infection by spores coming from infected *Berberis* could be ignored. Results for races 15B and 56 are given in Table 2.9. Race 15B became less frequent relative to race 56 as the season advanced and temperatures rose. The years concerned, 1951, 1952, and 1953, are relevant, because they probably identify race 15B as race 15B-1 and not 15B-1L which became abundant later. See the next section.

In the next chapter it is recommended that surveys be computerized. It would then be possible, almost automatically if the proper entries were made, to deter-

TABLE 2.8

The Effect of Temperature on the Percentage of Infection by a Culture of Race 15B-1 (Can.) of *Puccinia graminis tritici* Grown for 14 Uredial Generations on Two Susceptible Varieties of Wheat Mixed and in Competition with a Culture of Race 56[a]

	Temperature (°C)	
Generation	15	20
0[b]	50	50
1–2[c]	46	35
3–4[c]	28	35
5–6[c]	22	19
7–8[c]	21	3
9–10[c]	15	1
11–12[c]	8	0
13–14[c]	1[d]	0

[a]Data of Katsuya and Green (1967) for light infections. The data are independent of those in Table 2.7.

[b]Data adjusted for an equal (50%) start at generation 0. See the corresponding footnote to Table 2.7.

[c]The data for two successive generations and two varieties, Little Club and Red Bobs, are averaged.

[d] Note that, in contrast to the results in Table 2.7, race 15B-1 was competitively less fit than race 56 even at 15°C.

mine the effect of temperature on changes of gene frequency as the epidemic progressed. Evidence obtained in this way would be much superior to evidence obtained from artifacts; and it is badly needed. Temperature we believe to be very important as the key to the thermodynamics and energy relations in plant pathogens (see Chapter 6). It has been a grossly neglected factor.

TABLE 2.9

Number of Single-Pustule Cultures of Races 15B and 56 of *Puccinia graminis tritici*[a]

	Race			
Collection	56	15B	56	15B[b]
Early	15	11	100	73.3
Late	38	15	100	39.5

[a] Isolated from collections made early and late in 1951, 1952, and 1953 in fields of wheat to which the pathogen had spread from infected *Berberis* bushes. Extracted from the data of Roane *et al.* (1960).

[b] Calculated after adjusting race 56 to 100.

2.6 LOSS OF VIRULENCE ASSOCIATIONS IN RACE 15B-1

The mystery is, why did race 15B-1 ever become dominant? The answer usually given, that it could attack wheat varieties resistant to race 56, is incomplete. It did not need all the deleterious virulence associations to do so.

Though we can cannot solve the mystery, we can at least say that the dominance of race 15B-1 was relatively short-lived, from 1950 to 1955 (Green, 1971b). Later, race 15B-1L became abundant, and nowadays isolates of this race far outnumber those of race 15B-1. The point here is that the change from race 15B-1 to 15B-1L was accompanied by the loss of virulence for genes *Sr9a, Sr9b,* and *Sr15*. That is, the deleterious association of these virulences with virulence for genes *Sr9d* and *Sr9e* was shed in the change to race 15B-1L. At least the change from race 15B-1 to 15B-1L is understandable, and conforms with the other evidence.

In passing, it might astonish the uninitiated that two such very different races as 15B-1 to 15B-1L should fall together taxonomically. It should be explained that the key to "standard" race 15 involves the behavior of the fungus on 12 "standard" differential wheat varieties, none of which, except Vernal emmer which has gene *Sr9e,* seems to bear on what we have been discussing and think to be relevant.

2.7 THE DISSOCIATION OF VIRULENCE FOR GENES *Sr6* AND *Sr9e*

In Canada virulence for gene *Sr9e* behaves like that for gene *Sr9d;* it dissociates from virulence for *Sr6*. Records of virulence for gene *Sr9e* start more recently than those for gene *Sr9d,* and are therefore less complete. But the evidence is not in doubt; associated virulence is rare or absent.

Table 2.10 is constructed like Table 2.1. Thus, in 1974 83.7% of the isolates were virulent for gene *Sr9e,* and 11.1% for gene *Sr6*. If virulence had been randomly distributed in the population, 9.3% of the isolates would have had associated virulence for these two genes. In fact, none was found; dissociation was complete. Only in 1975 was associated virulence found, and then with very small frequency. It will be remembered that 1975 was the exceptional year when associated virulence for genes *Sr6* and *Sr9d* was also found; and the associated virulence for genes *Sr6* and *Sr9d* and for *Sr6* and *Sr9e* occurred in the same isolates.

In the United States virulence for gene *Sr9e* dissociates from virulence for gene *Sr6*. The dissociation that occurs only north of the Canadian border when virulence for gene *Sr9d* is involved with virulence for gene *Sr6* occurs on both sides of the border when virulence for *Sr9e* is involved (see Table 2.11). The United States survey also tests samples from Mexico. In Mexico, as in the United

TABLE 2.10

The Percentage of Isolates of *Puccinia graminis tritici* in Canada Virulent for Wheat Stem Rust Resistance Genes *Sr6* and *Sr9e*, Singly and in Combination, in the Years 1974–1978

	Year				
Resistance gene(s)	1974	1975	1976	1977	1978
Sr6[a]	11.1	14.2	5.3	23.1	17.1
Sr9e[a]	83.7	84.0	93.7	72.6	74.1
Sr6 + *Sr9e* expected[b]	9.3	11.9	5.0	16.8	12.7
Sr6 + *Sr9e* found[a]	0	0.9	0	0	0

[a] Data of Green (1975, 1976a,b, 1978, 1979).

[b] Expected on the assumption that the combination of virulences was selectively neutral and random.

States and Canada, virulence for genes *Sr6* and *Sr9e* is dissociated. Dissociation, it seems, is a phenomenon of the whole of North America. In Australia, too, virulence for genes *Sr6* and *Sr9e* is dissociated. Table 2.12 brings together information given by Luig and Watson (1970) for the seasons 1961/1962 through 1968/1969. Of a total of 5681 isolates, 976 were virulent for gene *Sr6* and 63 for gene *Sr9e*. The expected number of isolates virulent for both genes, i.e., with associated virulence, was $63 \times 976/5681 = 10.5$. None was found.

Evidence for the dissociation of virulence for genes *Sr6* and *Sr9e* has survived the vicissitudes of three separate surveys: those of Canada, the United States, and Australia. It must be strong. The practical implications are obvious. Any wheat

TABLE 2.11

The Percentage of Isolates of *Puccinia graminis tritici* in the United States Virulent for the Wheat Stem Rust Resistance Genes *Sr6* and *Sr9e*, Singly and in Combination, in the Years 1974–1978

	Years				
Resistance gene(s)	1974	1975	1976	1977	1978
Sr6[a]	6	11	7	24	18
Sr9e[a]	68	77	86	61	71
Sr6 + *Sr9e* expected[b]	4	8	6	15	13
Sr6 + *Sr9e* found[a]	0	0	0	0	0

[a] Data of Roelfs and McVey (1975, 1976) and Roelfs *et al.* (1977a, 1978b, 1979a).

[b] Expected on the assumption that the combination of virulences was selectively neutral and random.

TABLE 2.12

The Number of Isolates of *Puccinia graminis tritici* in Australia in the Years 1961/1962 through 1968/1969 against Which the Wheat Stem Rust Resistance Genes *Sr6* and *Sr9e* Were Effective or Ineffective[a]

Effective genes	Ineffective genes	Number of isolates
Sr6	*Sr9e*	63[b]
Sr9e	*Sr6*	976
Sr6, Sr9e		4822
	Sr6, Sr9e	0

[a] From the data of Luig and Watson (1970), in their Diagram 4 and Table 1.
[b] Australian strains 40-2; 40-2, 4, 5; 116-2; 116-4, 5; and 116-2, 3, 7.

variety with the two genes *Sr6* and *Sr9e* would be resistant to stem rust in North America and Australia, even though varieties with either of these genes singly would often be susceptible.

The question arises, why does virulence for gene *Sr9d* differ so widely from virulence for gene *Sr9e* in its behavior toward virulence for gene *Sr6* in the United States? Under the conditions represented in the United States survey, virulence for gene *Sr6* is fit to survive when it is associated with virulence for gene *Sr9d* but not when it is associated with virulence for gene *Sr9e*. On the previous evidence about temperature, it would seem that dissociation occurs at lower temperatures when virulence for gene *Sr9e* is involved. Virulence for gene *Sr6*, it would seem, is fit to survive the winter, i.e., to survive from year to year, in association with virulence for gene *Sr9d* but not *Sr9e*. The temperatures involved are not just those of a single season. They are those that prevail year after year, with warm months given more weight than cold months, because reproduction is quicker. By intrapolation using the data of Stakman and Harrar (1957), the latent period of wheat stem rust (the period from infection to sporulation) in 25 days at 5°C, 16 days at 10°C, 10 days at 15°C, and 8 days at 20°C.

2.8 THE CONCEPT OF FITNESS

Darwin defined fitness as success in leaving progeny, and this definition is generally accepted in population genetics. Thus, to judge by the current disease surveys, associated virulence for genes *Sr6* and *Sr9d* is fit to survive south of the United States–Canada border. Individuals there are successful in leaving progeny. North of the border, they are not; with rare exceptions, they are unfit.

Fitness is an attribute of the individual, which for fungi reproducing asexually we can interpret as the clone. All factor affecting success in leaving progeny affect fitness: the genotypes of host and pathogen, and the whole environment, physical and biotic, past and present, in every relevant detail.

To express fitness quantitatively, we shall in a later chapter and in a different context consider the progeny/parent ratio, and use this ratio as an important parameter in epidemiology. Here it is enough to note that, to survive passage across the Canadian border or to maintain itself in North America from year to year, the progeny/parent ratio must exceed 1. The more it exceeds 1, the fitter is the clone to survive and the greater its frequency relative to other clones.

2.9 THE DISSOCIATION OF VIRULENCE FOR GENE *Sr9e* FROM VIRULENCE FOR GENES *Sr9a, Sr9b,* AND *Sr15*

Virulence for genes *Sr9a*, *Sr9b*, and *Sr15* behaves like virulence for gene *Sr6*. As a rule, it dissociates from virulence for gene *Sr9e* throughout the whole of North America. Details are given in Table 2.13 for Canada and Table 2.14 for the United States.

TABLE 2.13

The Percentage of Isolates of *Puccinia graminis tritici* in Canada Virulent for Wheat Stem Rust Resistance Genes *Sr9e, Sr6, Sr9a, Sr9b,* and *Sr15,* Singly and in Combination, in the Years 1974–1977[a]

	Year			
Resistance gene(s)	1974	1975	1976	1977
Sr9e	83.7	84.0	93.7	72.6
Sr9a	8.7	6.9	5.6	5.2
Sr9b	14.8	6.4	6.6	25.2
Sr15	17.8	17.5	7.0	28.0
Sr9e + *Sr9a* expected[b]	7.3	5.8	5.2	3.8
Sr9e + *Sr9a* found	0.5	1.5	0	0.6
Sr9e + *Sr9b* expected[b]	12.3	5.4	6.2	18.3
Sr9e + *Sr9b* found	0.5	1.5	0	0.6
Sr9e + *Sr15* expected[b]	14.9	14.7	6.6	20.3
Sr9e + *Sr15* found	1.6	1.5	0.2	0.8

[a] Data of Green (1975, 1976a,b, 1978).

[b] Expected on the assumption that the combination of virulences was selectively neutral and random.

TABLE 2.14

The Percentage of Isolates of *Puccinia graminis tritici* in the United States Virulent for the Wheat Stem Rust Resistance Genes *Sr9e, Sr6, Sr9a, Sr9b,* and *Sr15,* Singly and in Combination, in the Years 1974–1977[a]

Resistance gene(s)	Year			
	1974	1975	1976	1977
Sr9e	68	77	86	61
Sr9a	22	14	10	18
Sr9b	4	12	7	26
Sr15	26	23	14	39
Sr9e + *Sr9a* expected[b]	15	11	9	11
Sr9e and *Sr9a* found	0	0.1	0	0
Sr9e and *Sr9b* expected[b]	3	9	6	16
Sr9e and *Sr9b* found	0	0	0	0
Sr9e and *Sr15* expected[b]	18	18	12	24
Sr9e and *Sr15* found	0	0	0	0

[a] Data of Roelfs and McVey (1975, 1976) and Roelfs *et al.* (1977a, 1978b). Isolates of formulas making up less than 0.6% of the total were not fully analyzed or recorded.

[b] Expected on the assumption that the combination of virulences was selectively neutral and random.

2.10 MATCHING VIRULENCE AND THE ABC–XYZ SYSTEM

Surveys have been reported in terms of a large number of races, and this has tended to confuse the central issues. The forest has not been seen for the trees.

In Canada, sampled during the years 1974 through 1977, two combinations of avirulence/virulence dominated the population of *P. graminis tritici.* Out of a total of 2017 isolates, 82% were avirulent for genes *Sr6, Sr9b* and *Sr15,* but virulent for *Sr9d* and *Sr9e;* and 13% reversed this order, being avirulent for genes *Sr9d* and *Sr9e,* but virulent for *Sr6, Sr9b,* and *Sr15.* Only the remaining 5% did not conform with this pattern. Details, extended to include gene *Sr9a,* are given in Table 2.15.

Similar data, less tightly patterned, are given for the United States and Mexico in Tables 2.16 and 2.17, respectively. Here again we see dissociation of virulence reflected. Virulence for genes *Sr9e* and *SrTmp,* on the one hand, and virulence for genes *Sr6, Sr9a, Sr9b,* and *Sr15,* on the other, are dissociated.

The gene *SrTmp,* not used in the Canadian surveys, occurs in the wheat variety Triumph 64 and is widespread among the hard red winter wheats (Roelfs and Mc Vey, 1975). Virulence for gene *SrTmp* is very closely associated with virulence for gene *Sr9e* in the United States and Canada (Roelfs *et al.,* 1979a).

TABLE 2.15

The Reaction of Isolates of *Puccinia graminis tritici* for the Wheat Stem Rust Resistance Genes *Sr6, Sr9a, Sr9b, Sr9d, Sr9e,* and *Sr15* in Canada during the Years 1974–1977[a]

Effective genes	Ineffective genes	Number of isolates[b]	Percent of total isolates
Sr6, 9a, 9b, 15	*Sr9d, 9e*	1652[c]	81.9
Sr9a, 9d, 9e	*Sr6, 9b,* 15	200[d]	9.9
Sr9d, 9e	*Sr6, 9a, 9b,* 15	63[e]	3.1
Other			5.1[f]

[a] Data of Green (1975, 1976a,b, 1978). Only isolates tested against all six relevant *Sr* genes are included. Gene *SrTmp* (see the Tables 2.16 and 2.17) was not used in the Canadian surveys; for information from outside see Table 2.18.
[b] The total number of isolates tested in the 4 years was 2017.
[c] Canadian formulas C18, C33, C42, C46, C49, C53, and C66.
[d] Canadian formulas C22, C25, and C57.
[e] Canadian formulas C35, C41, C59, and C63.
[f] No other avirulence/virulence combination on these six *Sr* genes exceeded 1% of the total.

TABLE 2.16

The Reaction of Isolates of *Puccinia graminis tritici* for the Wheat Stem Rust Resistance Genes *Sr6, Sr9a, Sr9b, Sr9e, Sr15,* and *SrTmp* in the United States during the Years 1974–1977[a]

Effective genes	Ineffective genes	Percent of total isolates
Sr6, 9a, 9b, 15	*Sr9e, Tmp*	73[b]
Sr9e, Tmp, 6, 9b	*Sr9a, 15*	10[c]
Sr9e, Tmp, 9a	*Sr6, 9b, 15*	8[d]
Sr9e, Tmp	*Sr6, 9a, 9b, 15*	3[e]
Sr9e, Tmp, 6	*Sr9a, 9b, 15*	1[f]
Other		5

[a] Data of Roelfs and McVey (1975, 1976) and Roelfs *et al.* (1977a, 1978b). The percentages are unweighted averages.
[b] United States formulas TBM, TDM, TLM, and TNM.
[c] United States formulas QCB, QCC, and QFB.
[d] United States formula QSH.
[e] United States formulas RKQ and RTQ.
[f] United States formulas RCR and RPQ.

TABLE 2.17

The Reaction of Isolates of *Puccinia graminis tritici* for the Wheat Stem Rust Resistance Genes *Sr6, Sr9a, Sr9b, Sr9e, Sr15,* and *SrTmp* in Mexico in the Years 1975 and 1976[a]

Effective genes	Ineffective genes	Percent of total isolates
Sr6, 9a, 9b, 15	*Sr9e, Tmp*	5[b]
Sr9e, Tmp	*Sr6, 9a, 9b, 15*	61[c]
Sr9e, Tmp, 9a	*Sr6, 9b, 15*	12[d]
Sr9e, Tmp, 6, 9b	*Sr9a, 15*	12[e]
Sr9e, Tmp, 6	*Sr9a, 9b, 15*	4[f]
Other		6

[a] Data of Roelfs and McVey (1976) and Roelfs *et al.* (1977a). 1975 and 1976 were the only two years for which adequate information was available. The total number of isolates was 281.

[b] United States formulas TDM, TLM, and TNM. These were present only in 1976.

[c] United States formulas RKQ and RTQ.

[d] United States formula QSH.

[e] United States formulas QCB and QFB.

[f] United States formulas RCR and RPQ.

We can divide the *Sr* genes into two groups, an ABC group and an XYZ group, with some (on present evidence) seemingly neutral genes as well. Genes *Sr6, Sr9a, Sr9b,* and *Sr15* belong to the ABC group, and genes *Sr9d, Sr9e,* and *SrTmp* to the XYZ group. In a population of *P. graminis tritici* virulence for any member of the ABC group tends to repel, and be dissociated from, virulence for any member of the XYZ group. The strength of the dissociation is conditioned by the enviroment, probably temperature; and, for example, gene *Sr9d* is clearly a member of the XYZ group in Canada but (on available evidence) neutral, or nearly neutral, in the United States, a matter discussed in Section 2.5.

We introduce the concept of matching genes. A wheat variety with two matched genes, one from the ABC and the other from the XYZ group, is likely to be resistant to a higher proportion of the *P. graminis* population than a variety with two unmatched genes both from the ABC group or both from XYZ group. This can be seen at a glance at Tables 2.15, 2.16, or 2.17. Matching is important in wheat breeding, and will be discussed in more detail later.

Plant pathologists are familiar with mating types in fungi. Matching types, recognized in a very different context and probably with a very different biochemical background, are another concept we shall have to become familiar with.

Virulences within the ABC group or within the XYZ group tend to associate. The same biochemical processes that cause virulences from different groups to dissociate drive virulences within the groups to associate. Again, a quick look at Table 2.15 shows how virulences on genes *Sr6*, *Sr9b*, and *Sr15*, within the ABC group, tend to associate, and this Table 2.15 together with Tables 2.16 and 2.17 show association between virulences on genes *Sr9d*, *Sr9e*, and *SrTmp* within the XYZ group.

The opposing processes of virulence dissociation and association are neatly summarized in Table 2.18 compiled from the survey of Roelfs *et al.* (1979a). The table gives data for the United States as a whole, then for the different ecological areas within the United States, and finally for Mexico and Canada. Some of the samples were small and separately carry little weight, but together the samples illustrate our topic vividly; the table should be taken to mean that 790 isolates of *P. graminis tritici* of very diverse origin all followed practically the same pattern. Dissociation between virulence for gene *Sr15* from the ABC group and for genes *Sr9e* or *SrTmp* from the XYZ group is practically complete; the percentage virulent for gene *Sr15* plus the percentage virulent for gene *Sr9e* adds up to 100%. There was no associated virulence between the two groups, nor was

TABLE 2.18

The Percentage of Isolates of *Puccinia graminis tritici* in North America Virulent for the Wheat Stem Rust Resistance Genes *Sr15, Sr9e,* and *SrTmp* during 1978[a]

Area[b]	*Sr15*	*Sr9e*	*SrTmp*
United States	29	71	72
1S	80	20	20
1N	55	45	48
2	82	18	18
3	26	74	75
5	19	81	81
6	26	74	74
7	40	60	60
8	0	100	100
Mexico	100	0	0
Canada	0	100	100

[a] Data of Roelfs *et al.* (1979a). The number of isolates was 790.

[b] Area 1S, Texas south of 30°; area 1N, central Texas; area 2, south and east of north central Texas, Arkansas, southern Missouri and Illinois, lower Michigan, Ohio, western West Virginia, and North Carolina; are 3, north central Texas to the Platte River in Nebraska; area 5, eastern three-fourths of Minnesota, Iowa, northeastern Missouri, and northwestern Illinois and Wisconsin; area 6, northeastern Montana, North Dakota, eastern South Dakota, and the western fourth of Minnesota; area 7, from the Nebraska panhandle through central Montana; area 8, the Western States. No data were published for area 4. The only isolates out of step belonged to the rare race MBC, virulent for genes *Sr15* and *SrTmp* and avirulent for gene *Sr9e*.

there associated avirulence. Equally conspicuous in the table is the strong association between virulences for genes *Sr9e* and *SrTmp*.

Virulence, not avirulence, dictates the pattern. There are indeed avirulence associations, but only as a consequence. Thus, an isolate virulent for the gene *Sr9e* is likely to be avirulent for genes *Sr6, Sr9a, Sr9b,* and *Sr15,* giving rise to secondary associations of avirulence. Moreover, avirulences can associate even when they are for genes of both the ABC and XYZ groups. Race 56, very frequent in population of *P. graminis tritici* in North America between 1935 and 1950, is avirulent for *Sr6, Sr9a, Sr9b, Sr9d,* and *Sr9e*. Avirulence frequencies are being reduced as wheat breeders use more and more *Sr* genes; this is a topic of Chapter 4. We need only note the self-evident proposition that avirulence must initially have been frequent for all *Sr* genes that have since been used in breeding wheat for stem rust resistance; otherwise the genes would not have been used in breeding for resistance.

2.11 GENES *Sr7b, Sr10, Sr11,* AND *SrTt1* IN THE XYZ GROUP

Genes *Sr7b, Sr10, Sr11,* and *SrTt1,* like genes *Sr9d, Sr9e,* and *SrTmp,* belong to the XYZ group. Virulence for these genes dissociates from virulence for genes of the ABC group, like *Sr6*. Their membership of the XYZ group is not so tight as that of genes *Sr9e* and *SrTmp,* and is not even evident every year; there is a seasonal effect which might reflect climatic differences between the years or just dates of sampling.

Evidence about genes *Sr10* and *Sr11* is available for many years, and we begin with it. Table 2.19 summarizes the data of 8 years in Canada about virulence and avirulence for genes *Sr6* (in the ABC group) and *Sr11*. The two rows at the bottom show that associated virulence on genes *Sr6* and *Sr11* was considerably and significantly less than expected in 5 of the 8 years. In these years gene *Sr11* clearly belongs to the XYZ group.

This finding has an echo in an observation of Luig (1979). He remarked that for nearly a decade (i.e., prior to 1979) all races prevalent in Australia possessed virulence on plants with gene *Sr11*. Because the leading Australian wheat cultivars (Pinnacle, Olympia, Insignia, Falcon) do not possess the gene *Sr11,* it appears that in the present Australian rust flora the capacity to attack plants with gene *Sr11* is no longer associated with loss of aggressiveness. This is in contrast to the capacity of the rust flora before 1954. At this earlier date, despite large areas sown to Gabo and other cultivars with gene *Sr11,* races avirulent on plants with this gene constituted a major part of the rust flora. In the terminology of the next section, gene *Sr11* is now weak but was strong before 1954. Table 2.19 suggests the explanation. Before 1954, and especially between 1945 and 1954, virulence for gene *Sr6* was common, because the cultivar Eureka, which pos-

TABLE 2.19

The Number of Isolates of *Puccinia graminis tritici* in Canada in the Years 1970–1977 against Which the Wheat Stem Rust Resistance Genes *Sr6* and *Sr11* Were Effective or Ineffective[a]

Effective genes	Ineffective genes	1970	1971	1972	1973	1974	1975	1976	1977
Sr6	*Sr11*	176	66	204	84	347	259	485	400
Sr11	*Sr6*	17	26	40	11	15	9	25	4
Sr6, 11		9	9	30	10	34	26	85	103
	Sr6, 11 expected[b]	16	18	35	10	43	42	26	125
	Sr6, 11 found[a]	1	1	7	1	33	38	7	146

[a] Data of Green (1971a, 1972a,b, 1974, 1975, 1976a,b, 1978).

[b] Expected in the assumption that the combination of virulences was selectively neutral and random.

sessed gene *Sr6,* was the dominant cultivar until Gabo was released. Gabo with gene *Sr11* was released in a background of virulence for gene *Sr6,* and, in line with the results in Table 2.19, virulence for gene *Sr11* was depressed. Nowadays, virulence for gene *Sr6* is infrequent in Australia, and depression of virulence for gene *Sr11* by virulence for gene *Sr6* is insignificant. The same evidence explains a confusion in the literature. Vanderplank (1968) classed gene *Sr11* as strong, on the evidence of the early success of Gabo in Australia. Osoro and Green (1976) disputed this on evidence that, in the isolates of *P. graminis tritici* they tested, virulence for gene *Sr11* did not depress fitness. The point about their evidence is that all the isolates they tested were avirulent for gene *Sr6;* stabilizing selection through dissociation of virulences for *Sr6* and *Sr11* was necessarily absent.

Table 2.20, reporting data comparable with those in the bottom two rows of Table 2.19, confirms that genes *Sr10* and *Sr11* belong to the XYZ group. Dissociation occurs between virulence for genes *Sr10* and *Sr11* on the one hand and for genes of the ABC group on the other. Also brought out in Tables 2.19 and 2.20 is the regularity of the environmental effect on dissociation, as measured by yearly differences. The years 1970, 1971, 1972, 1973, and 1976 were years of considerable dissociation, and this holds generally for all gene pairs. The years 1974, 1975, and 1977 were years of little or no dissociation, and this too holds generally for all gene pairs. Environmental effects, whatever they might have been, were consistent. Identifying the source of the environmental effects is one of the major problems to be solved.

The wheat stem rust resistance gene *Sr7b* came into the Canadian survey in 1978 (Green, 1979). Virulence for gene *Sr7b* dissociates from virulence for

TABLE 2.20

The Number of Isolates, Expected and Found, of *Puccinia graminis tritici* Virulent in Canada for Wheat Stem Rust Resistance Genes *Sr6, Sr9a, Sr10, Sr11,* and *Sr15* in Appropriate Pairs from 1970 to 1977

	Year							
Sr genes	1970	1971	1972	1973	1974	1975	1976	1977
6 and *10* expected[b]	17	19	35	10	32	46	31	145
6 and *10* found[a]	2	0	3	0	31	40	6	143
9a and *10* expected[b]	21	27	18	21	33	22	31	34
9a and *10* found[a]	3	0	0	0	8	14	5	17
15 and *10* expected[b]	35	28	42	10	69	56	38	177
15 and *10* found[a]	18	9	11	2	47	49	13	162
9a and *11* expected[b]	20	19	36	11	27	20	26	31
9a and *11* found[a]	6	1	4	2	10	9	4	10
15 and *11* expected[b]	34	28	42	11	66	51	34	153
15 and *11* found[a]	12	7	12	3	44	37	9	154

[a, b] These correspond with footnotes *a* and *b* of Table 2.19, and the data in Table 2.20 correspond with those in the last two rows of Table 2.19.

genes *Sr6, Sr9a, Sr9b,* and *Sr15*. Evidence about virulence for genes *Sr6* and *Sr7b* is given in Table 2.21. The expected number of isolates virulent for both genes, if the distribution was random, was 37; the number found was 3. With virulence for genes *Sr7b* and *Sr9a* the numbers expected and found were 20 and 1, respectively; for genes *Sr7b* and *Sr9b,* 37 and 2; and for genes *Sr7b* and *Sr15,* 47 and 3.

TABLE 2.21

The Number of Isolates of *Puccinia graminis tritici* in Canada in 1978 against Which the Wheat Stem Rust Resistance Genes *Sr6* and *Sr7b* Were Effective or Ineffective[a]

Effective gene(s)	Ineffective gene(s)	Number of isolates
Sr6, 7b		37
Sr6	*Sr7b*	215
Sr7b	*Sr6*	49
	Sr6, 7b expected[b]	37
	Sr6, 7b found[a]	3

[a] Data of Green (1979).

[b] Expected on the assumption that the association of virulences is selectively neutral and random.

In the United States there is also some evidence for the dissociation of virulence for gene *Sr7b* from virulence for members of the ABC group of genes. It is less clear than in Canada.

The gene *Sr7b* has long been used in North America. It is one of the genes of the Canadian cultivar Marquis (McIntosh, 1973) which dominated the spring-wheat fields of North America early this century. It is interesting to speculate that Marquis missed by one gene being substantially resistant to stem rust. If besides gene *Sr7b* Marquis had had one of the genes *Sr6, Sr9a, Sr9b,* or *Sr15,* the great stem rust epidemic of 1916 would possibly not have occurred. Marquis was crossed with Jaroslav emmer to give the lines Hope and H-44, which played a prominent part in wheat breeding against disease. In retrospect it can be seen as unfortunate that this cross combined genes *Sr7b* and *Sr9d,* a combination carried over to the cultivars Renown and Redman. These genes both belong to the XYZ group, and virulence for them is highly associated; of the 218 isolates in the 1978 Canadian survey virulent for gene *Sr7b,* 215 or 98.6% were also virulent for gene *Sr9d*. The two genes together would give practically no better protection against stem rust than either of them singly. (There are other genes in Hope and H-44, namely *Sr2* and *Sr17,* which presumably helped to establish their reputation for stem rust resistance. There is no evidence at all about the affinity of gene *Sr2* for the ABC or XYZ groups, and *Sr17* seems to be more or less neutral.)

In Australia virulence for gene *Sr7b* has been associated with virulence for genes *Sr6* and *Sr15,* as in Australian races 126-1, 6, 7 and 222-1, 2, 6, 7 (virulent for gene *Sr7b* and both *Sr6* and *Sr15*). These four races were predominant until 1955 (Watson, 1958). What is also remarkable is that the introduction of the wheat cultivar Eureka, with gene *Sr6,* did not substantially alter the standard races 126 and 222, both of which are virulent for gene *Sr7b,* even when the introduction of virulence for gene *Sr6* into the fungus population inevitably followed. (Marquis wheat is one of the 12 differential varieties for determining "standard" stem rust races; something is therefore known about virulence for gene *Sr7b* in the early days.)

Information about virulence for gene *SrTt1* has appeared in the Canadian surveys since 1975 (Green, 1976a,b, 1978, 1979). The gene loosely belongs to the XYZ group. To give information in relation to gene *Sr6* only, during the four years 1975 through 1978 the number of isolates found to be virulent for both genes *Sr6* and *SrTt1* was 45, compared with 233, the number expected if the virulence were randomly distributed.

To summarize, genes *Sr6, Sr9a, Sr9b,* and *Sr15* belong to the ABC group, and *Sr7a, Sr9d, Sr9e, Sr10, Sr11, SrTmp,* and *SrTt1* to the XYZ group. Many belong loosely, according to the available surveys. It is possible that some of the genes that now seem to be neutral will be assigned to one or other group when environmental factors are better understood.

2.12 VIRULENCE DISSOCIATION AND STABILIZING SELECTION

The adaptation of the pathogen to the host is a topic of Chapter 4, but some aspects are more conveniently dealt with here.

Adaptation is a rule of nature. In the present context, the population of the pathogen tends to adapt itself to the population of the host; it must adapt to survive. In different terms, there is directional selection in favor of virulence in the pathogen. But there is a constraint on directional selection. The previous sections have dealt with such a constraint. If a wheat cultivar had two *Sr* genes, directional selection in *P. graminis tritici* in favor of associated virulence for these genes would be constrained to a greater or lesser extent by the tendency of virulence to dissociate, provided that the two *Sr* genes came one from each of the ABC and XYZ groups. This constraint is called stabilizing selection or homeostasis. It is the opposite of directional selection.

Stabilizing selection via the tendency for virulence to dissociate has three effects that are best considered separately in the next three section.

2.13 STABILIZING SELECTION IN VERTICAL RESISTANCE

Suppose (to make the example concrete) a wheat cultivar with genes *Sr6* and *Sr9e* were introduced into North America. The North America population of *P. graminis tritici* has at present almost infinitesimally little associated virulence for genes *Sr6* and *Sr9e,* found recently only in the rare Canadian race C50. The wheat cultivar would thus have great vertical resistance, stemming from the tendency of the virulences for genes *Sr6* and *Sr9e* to dissociate.

One notes here that genes give vertical resistance only against inoculum coming from host plants with other genes (Vanderplank, 1963, pp. 197–198). A wheat cultivar with the genes *Sr6* and *Sr9e* is vertically resistant only to inoculum coming from infected plants without these two genes. Vertical resistance therefore results from stabilizing selection in populations of *P. graminis tritici* on plants without the resistance genes, i.e., without *Sr6* and *Sr9e* in combination. The stabilizing selection results from dissociation of virulence in what Robinson (1979, 1980) conveniently calls the exodemic.

Vertical resistance against wheat stem rust can be expected whenever resistance genes are drawn from both the ABC and the XYZ groups in appropriate environments. The stability of the resistance is likely to depend on the selection of genes strongly belonging to one or other of the two groups in the appropriate environment. Thus, as representative of the XYZ group, gene *Sr9e* is likely to be better than gene *Sr9d* in the environment sampled by the United States survey, but not necessarily better in the environment sampled by the Canadian survey.

Although this section has been written around stabilizing selection arising from the dissociation of virulences, i.e., although it has been written around vertical resistance given by a combination of two or more resistance genes, it is not implied that this is the only source of stabilizing selection. There could in theory be stabilizing selection in vertical resistance given by a single gene, but this is not part of the story of this chapter.

2.14 STABILIZING SELECTION AND THE HORIZONTAL RESISTANCE EQUIVALENT

Let us continue with the same example for illustration. Suppose that agronomically adapted wheat cultivars with genes *Sr6* and *Sr9e* became widely grown in North America and that races like C50 virulent for both these genes became the common races. The vertical resistance given by the genes would be largely nullified. But that would not be the end of the story. The races virulent for the genes would be less aggressive and therefore less destructive. This is a clear implication of the rarity of these races. There is also direct evidence. In Canada Katsuya and Green (1967) and Green (1971b) observed that the few isolates of race C50 (= race 15B-5) which they found were unaggressive and could not maintain themselves in nature.

Lack of aggressiveness in the pathogen is the counterpart and equivalent of horizontal resistance in the host. We therefore use the term horizontal resistance equivalent. The distinction between horizontal resistance and the horizontal resistance equivalent is wide. Horizontal resistance is an attribute of the host operative against all isolates of the pathogen. The horizontal resistance equivalent is an attribute of the pathogen operative against all varieties of the host. In epidemiological effect they may be the same, but their origin is in different genomes.

The narrow distinction is between the horizontal resistance equivalent and other forms of reduced aggressiveness. Whereas there is no evidence at present to relate other forms of aggressiveness to known genes, the horizontal resistance equivalent is determined by the same genes in the pathogen as determine virulence, and can therefore be identified, via the gene-for-gene hypothesis, with resistance genes in the host. It is the host that the plant breeder manipulates, and the identification helps the breeder greatly.

The horizontal resistance equivalent, like horizontal resistance itself but unlike vertical resistance, does not need two host genotypes to be operative. It can operate in an esodemic. That is, if a variety with two genes like *Sr6* and *Sr9e* were introduced into cultivation, there would first be vertical resistance in the host during the initial exodemic, to be followed by a horizontal resistance equivalent during the subsequent esodemic.

A horizontal resistance equivalent can be expected whenever the host has resistance genes from both the ABC and XYZ groups, but not when the resistance genes all come from one group or the other.

A horizontal resistance equivalent in the pathogen, like horizontal resistance in the host, would slow an epidemic down, i.e., it would make for slow rusting. Matched genes from the ABC and XYZ groups would have a double action in reducing the level of disease reached at the end of the season. Through vertical resistance in the host they would delay the onset of the epidemic (exodemic); through a horizontal resistance equivalent in the pathogen they would slow down the epidemic (esodemic) after it had started.

How important is a horizontal resistance equivalent in the pathogen relative to horizontal resistance in the host? It depends on the diesease. In wheat stem rust the importance might be substantial. In potato blight, on the other hand, a horizontal resistance equivalent is probably absent or, at best, of little importance in most countries. Varieties of *Solanum tuberosum sensu stricto* have no *R* genes. Many of them, especially the late-maturing European varieties like Pimpernel and Capella, have very great horizontal resistance. But in the absence of *R* genes *Phytophthora infestans* attacking them can have no horizontal resistance equivalent. It remains to be shown whether the *R* genes introduced from *S. demissum* have brought a horizontal resistance equivalent into *P. infestans*. If they have, the amount must be small for various reasons, including the fact that most widely grown potato varieties (like Russet Burbank, Katahdin, or Sebago in North America) are still those without *R* genes. If a horizontal resistance equivalent is to be looked for in potatoes, it should be looked for in Central America where *Solanum* spp. with *R* genes are indigenous.

It must be emphasized that horizontal resistance in the host and a horizontal resistance equivalent in the pathogen, though they operate in the same direction, are different phenomena, probably with different biochemical backgrounds.

2.15 STABILIZING SELECTION INHIBITING EPIDEMICS

We have been discussing stabilizing selection which arises from the dissociation of virulences for genes drawn together from both the ABC and XYZ groups and promotes vertical resistance in the exodemic and a horizon resistance equivalent in the esodemic. This stabilizing selection can also have a third effect. It can inhibit an epidemic from starting at all; the vertical resistance and the horizontal resistance equivalent would then be irrelevant. Inhibition is clearly suggested by the almost complete absence in the North American and Australian surveys of associated virulence for genes *Sr6* and *Sr9e* and the rarity of virulence for genes *Sr9a, 9b,* or *15* in the ABC group associated with virulence for *Sr9e* in the XYZ group.

One must picture associated virulences as arising locally and sporadically, and then, if the pressure of stabilizing selection is great enough, dying out. Luig (1979) has given such a picture of events in Australia. Since 1961, he observed, each race of *P. graminis tritici* with newly acquired virulence to attack plants with the gene *Sr9e* has failed to establish itself and has disappeared the following season. The main associations involved here are between virulence for gene *Sr9e* and for gene *Sr9b* or *Sr15*. Evidently these races were unfit to survive; that is, the progeny/parent ratio, measured from season to season, did not exceed one.

The successful inhibition of epidemics, via the inhibition of all except sporadic virulence associations, can only be judged against the environmental background. An extent of inhibition adequate for the old established areas where wheat growing preceded the conscious breeding of resistant varieties may well prove inadequate if wheat culture were extended to less favored environments.

2.16 THE SECOND GENE-FOR-GENE HYPOTHESIS

Flor's gene-for-gene hypothesis, discussed in Chapter 6, is about gene identities. Resistance genes in the host identify virulence genes in the pathogen, and vice versa. The present chapter is essentially about gene qualities. Virulence for genes in the ABC group differs in quality from virulence for genes in the XYZ group; and these virulence differences in the pathogen inevitably reflect differences in the host genes themselves.

Quality differences were first recognized in the second gene-for-gene hypothesis formulated by Vanderplank (1975). In host–pathogen systems in which there is a gene-for-gene relationship, the quality of a resistance gene in the host determines the fitness of the matching virulence gene in the pathogen to survive when this virulence is unnecessary; and reciprocally the fitness of the virulence gene to survive when it is unnecessary determines the quality of the matching resistance gene as judged by the protection it can give to the host. Resistance genes are called strong or weak according to whether the matching virulences are infrequent or frequent in the pathogen population.

Since the hypothesis was formulated we have come to recognize that the strength of the resistance gene (as defined in the previous sentence) is not an inherent property but is influenced by the genetic background and the environment. Thus, in Canada gene *Sr6* is strong, and therefore useful to the wheat breeder, when it is coupled with gene *Sr9d* but weak when coupled with gene *Sr9a,* because virulence for genes *Sr6* and *Sr9d* tends to dissociate and virulence for genes *Sr6* and *Sr9a* tends to associate. Therefore, the associated virulence relevant to the hypothesis is infrequent for genes *Sr6* and *Sr9d* and frequent for genes *Sr6* and *Sr9a*. To illustrate the environmental effect, genes *Sr6* and *Sr9d,* coupled, are strong in the environment sampled by the Canadian surveys and weak in the environment sampled by the United States surveys.

The powerful effect of genetic background is well illustrated by the metamorphosis of gene *Sr9e*. In many countries virulence for this gene is common even though the gene itself is absent from the wheat fields. The gene, by definition, is weak. But when virulence for *Sr6* or other genes of the ABC group is present in *P. graminis tritici* virulence for genes *Sr9e* is suppressed. The gene *Sr9e* in the background of genes of the ABC group becomes strong.

In the sphere of practical agriculture, the second gene-for-gene hypothesis withdraws the emphasis from the genetics of the host plant and places it on the genetics of the pathogen. A wheat breeder manipulates the *Sr* genes in *Triticum* but whether the manipulation will be fruitful depends on the population genetics of *Puccinia*.

The sections that follow are concerned largely with the practical problems associated with breeding for stem rust resistance.

2.17 BREEDING WHEAT FOR STEM RUST RESISTANCE

From what has been written in this chapter it is clear that to breed wheat for resistance to stem rust at least two resistance genes should be used, at least one from each of the ABC and XYZ groups. Their belonging to these groups is as important as their number. Consider two Canadian wheat cultivars, on the information given by Green and Campbell (1979). Selkirk, licensed in 1953, has resistance to stem rust that (in Canada) has proved to be stable over the years. Canthatch, licensed in 1959, is now susceptible. Both have, as far as is known, five *Sr* genes. Selkirk has *Sr6, 7b, 9d, 17,* and *23*. Canthatch has *Sr5, 7a, 9g, 12,* and *16*. Of Selkirk's five genes, one, *Sr6,* belongs to the ABC group, and two, *Sr7b* and *9d,* to the XYZ group. Of Canthatch's five genes all seem to be neutral or nearly neutral, and none, on present evidence, belongs to either group.

Canthatch's failure as a resister of stem rust is a warning that the mere accumulation of resistance genes in a wheat variety is not the path to stable resistance.

Even worse than the random use of *Sr* genes would be the use of genes all from one group, whether it be the ABC or the XYZ. The tendency toward association of virulences for genes within a group is clearly brought out in Tables 2.15, 2.16, and 2.17.

2.18 MULTILINES AND MIXED VARIETIES

Instead of accumulating resistance genes in a variety, the genes can be dispersed in multilines and mixed varieties. Both multilines and mixed varieties have their advocates. We are here concerned with their use in relation to the population genetics of *P. graminis tritici,* but before we can discuss this we must digress awhile.

A multiline cultivar is a mechanical mixture of near-isogenic lines each of which contains a different gene for resistance. Jensen (1952) described the production of multilines in oats. Borlaug (1953, 1965) extended the concept to wheat for the control of stem rust. The major use so far has been against crown rust of oats, and success over an area of a million acres annually in Iowa and neighboring states has been reported by Frey *et al.* (1977). The principle involved is one of gene diversification, stressed from the start by Jensen (1952). In a field of wheat planted to different lines, a clone of *P. graminis tritici* able to attack one of the wheat lines would be obstructed in its dispersal by the presence of other wheat lines resistant to it. Line diversity would hinder the pathogen's spread. Even if clones of *P. graminis tritici* were present that could attack two or more lines, the hindrance would persist, to a greater or smaller extent, unless a superrace of *P. graminis tritici* developed that could attack all the lines.

The concept of multilines carries its own inherent contradiction. A multiline against, say, wheat stem rust, with each near-isogenic line carrying a different *Sr* gene, would, from the very definition of the word isogenic, be nearly genetically uniform in relation to leaf rust, stripe rust, powdery mildew, bunt, smut, and all the other diseases and pests of wheat. Multilines would in fact fall into the very trap of genetic uniformity which they are designed to avoid. If genetic diversity be a safeguard against epidemics of disease, then a multiline against one particular disease could promote epidemics against all other diseases and pests. Of course, one might argue that composite multilines should be developed as frequently and with as much background diversity as conventional cultivars. But with limited manpower and resources this is a counsel of perfection unlikely to be realized in practice; and one must concede the possibility that multilines, unless developed with less attention to isogenes, could do more harm than good.

Mixed varieties escape this criticism. Like multilines, they make a virtue of diversity. Unlike multilines, they pursue this virtue consistently. Mixed varieties are grown from mechanical mixtures of seed of several varieties each of which carries, say, a different *Sr* gene and in the background has the diversity of other genes associated with diverse varieties.

Like multilines, mixed varieties were first proposed for the control of oat disease. Soon after the great epidemic of victoria blight of oats caused by *Helminthosporium victoriae,* Rosen (1949) proposed the use of mixed varieties to control both victoria blight and crown rust. Because all known Victoria oat derivatives that were resistant to race 45 of *Puccinia coronata* were susceptible to *H. victoriae,* and most derivatives of the variety Bond were resistant to *H. victoriae* but susceptible to *P. coronata,* Rosen suggested using a varietal mixture of which about 70% of the plants were Victoria derivatives and 30% Bond derivatives. Recently the advantages of cereal variety mixtures have been extolled by Wolfe (1978), especially in relation to barley powdery mildew caused by *Erysiphe graminis hordei*.

We can now return to the population genetics of *P. graminis tritici* in so far as it concerns multilines and variety mixtures. It can immediately be seen that multilines or variety mixtures with *Sr* genes from the same group are practically worthless for the control of stem rust. Multlines or mixtures with the genes *Sr6, Sr9a, Sr9b,* and *Sr15,* all of the ABC group, would quickly succumb to a superrace against which all the genes would be ineffective. Clones of this superrace are already dominant in Mexico (see Table 2.17) and not rare in Canada and the United States (see Tables 2.15 and 2.16). Similarly a multiline or mixture with genes *Sr9d, Sr9e, Sr10, Sr11,* and *SrTmp,* all from the XYZ group, would quickly succumb to a superrace of associated virulences.

A multiline or mixture would have to have *Sr* genes from the two opposing groups. But how many would be needed? Plants in a 2-line or 2-variety mixture with genes *Sr6* and *Sr9e* would be susceptible to clones of *P. graminis tritici* virulent for one *Sr* gene or the other. Such clones are not infrequent. But in North America and Australia it would not easily be susceptible to a superrace virulent for both these genes. Would two lines or varieties in, say, a 50:50 mixture adequately hinder the spread of clones virulent on one or other gene? The answer depends on the circumstances. The hinderance would probably be adequate to relegate a mild stem rust epidemic to economic unimportance, but not a very severe epidemic. Could one improve performance by adding other genes from the ABC and XYZ groups? Would a multiline or mixture with nine genes, *Sr6, Sr9a, Sr9b, Sr9d, Sr9e, Sr10, Sr11, Sr15,* and *SrTmp,* be much safer than a multiline or mixture with *Sr6* and *Sr9e* only? One must doubt this, because all the plants with *Sr* genes in the ABC group would soon go down to clones of *P. graminis tritici* with associated virulence for them all, as would all the plants with *Sr* genes in the XYZ group. What would seem like a 9-line or a 9-variety mixture would soon be in effect no more than a 2-line or 2-variety mixture.

2.19 POSSIBLE SUPERGENES

The *Sr9* locus for reaction to *Puccinia graminis tritici* is on chromosome 2BL of wheat, and was first demonstrated by Green *et al.* (1960). Of the genes at or near the *Sr9* locus, genes *Sr9a* and *Sr9b* belong to the ABC group, and genes *Sr9d* and *Sr9e* to the XYZ group.

If gene *Sr9e* could be coupled with gene *Sr9a* or *Sr9b,* there would be a tightly linked supergene, which wheat breeders could mainpulate as if it were a single gene and which would be effective against almost all clones of *P. graminis tritici* now known in North America. A coupling of gene *Sr9d* with *Sr9a* or *Sr9b,* although less effective, would also be of considerable value in Canada.

The genes are allelic or are tightly linked, i.e., they are allelic or pseudoallelic (see Section 6.4). Could they be coupled by intragenic or intergenic recombination? There are precedents to suggest they could. There has been parallel recombination with flax rust caused by *Melampsora lini* (Flor, 1965), maize rust caused by *Puccinia sorghi* (Saxena and Hooker, 1968), and wheat leaf rust caused by *P. recondita* (Dyck and Samborski, 1970) (see Section 6.4). The prize of a supergene is perhaps the greatest prize available to wheat breeders battling with stem rust. Instead of the apparently unending quest for new *Sr* genes wheat breeders could rely on the stabilizing selection against virulence that supergenes ought to give. The screening of tens of thousands of plants in a search for suitable recombinants would be a small price to pay.

2.20 OAT STEM RUST

We turn now to *Puccina graminis avenae*. The evidence suggests the rudiments of an ABC and an XYZ group for this fungus too, with *Pg8* in the ABC and *Pg9* and *13* in the XYZ group.

There is clear evidence for dissociation of virulence for genes *Pg8* and *Pg9* in the data of the Canadian surveys from 1971 through 1978 (see Table 2.22). In the west of Canada virulence for gene *Pg8* in almost universal, and virulence for gene *Pg9* correspondingly rare. In the east the imbalance is not so marked. Data from the oat stem rust nurseries are available from 1974. They fully substantiate the evidence for dissociation.

TABLE 2.22

The Number of Isolates of *Puccinia graminis avenae* in Canada against Which the Oat Stem Rust Resistance Genes *Pg8* and *Pg9* Were Effective or Ineffective[a]

Effective gene(s)	Ineffective gene(s)	Western Canada[b]	Eastern Canada[c]	Nurseries[d]
Pg8	*Pg9*	14	103	117
Pg9	*Pg8*	1481	57	755
Pg8, 9		40	7	14
	Pg8, 9 expected[e]	17	36	100
	Pg8, 9 found	4	1	0

[a] Data of Martens and Anema (1972), Martens (1972, 1974, 1975), Martens and McKenzie (1976a,b) and Martens (1978, 1979).

[b] Manitoba, Saskatchewan, and Alberta, in the years 1971–1978.

[c] Nova Scotia, New Brunswick, Quebec, and Ontario, in the years 1971–1978.

[d] Data from oat stem rust trap nurseries, in the years 1974–1978.

[e] Expected on the assumption that the association of virulence is selectively neutral and random.

TABLE 2.23

The Number of Isolates of *Puccinia graminis avenae* for the Oat Stem Rust Resistance Genes *Pg8, Pg9,* and *Pg13* in Canada during the Years 1974–1978[a]

Effective genes	Ineffective genes	Number of isolates	Percent of total isolates
Pg9, 13	*Pg8*	966	90.9
Pg8, 9, 13		38	3.6
Pg8	*Pg9, 13*	35	3.3
Pg8, 13	*Pg9*	21	2.0
Pg8, 9	*Pg13*	2	0.2
Pg9	*Pg8, 13*	1	0.1[b]

[a] Data from the same sources as in Table 2.22. Only those isolates are included that were tested against all three genes.

[b] Race C29.

The gene *Pg13* belongs to the same group as gene *Pg9*. The data about gene *Pg13* in Canada do not go so far back as those about genes *Pg8* and *9*, but the analysis of more than a thousand isolates which had been tested against all three genes is conclusive. In only one isolate was virulence for genes *Pg8* and *13* associated (see Table 2.23).

In the United States survey the evidence is in the same direction, but is erratic. As in Canada, virulence for gene *Pg8* is prevalent in the west and that for genes *Pg9* and *13* correspondingly rare. This makes statistical analysis for any one year unsatisfactory. Data for the years 1971 through 1977 are pooled in Table 2.24. Dissociation of virulence for genes *Pg8* and *13* was complete, and that of

TABLE 2.24

The Reaction of Isolates of *Puccinia graminis avenae* for the Oat Stem Rust Resistance Genes *Pg8, Pg9,* and *Pg13* in the United States during the Years 1971–1978[a]

Effective gene(s)	Ineffective gene(s)	Average percent
Pg9, 13	*Pg8*	89
Pg8, 9, 13		4
Pg8, 13	*Pg9*	3
Pg8	*Pg9, 13*	1
Pg13	*Pg8, 9*	1
Unspecified		2[b]

[a] Data of Roelfs and Rothman (1972, 1973, 1974, 1975, 1976), Roelfs *et al.* (1977b, 1978a, 1979b). The total number of isolates in the 8 years was 9631.

[b] The surveys omit numerical details for races accounting for less than 0.6% of the total.

virulence for genes *Pg8* and *9* almost complete. But a year earlier, in 1970, race 94 with virulence for genes *Pg8* and *9* (but not *13*) was abundant. As one would expect, this abundance was in the Northeastern states (Roelfs and Rothman, 1971).

Data for other continents have been compiled by Martens *et al.* (1976). Associated virulence for genes *Pg8* and *9* was not found in Austria, Czechoslovakia, and Romania, but was common in Ethiopia and Kenya. In these East African countries oats are grown only in cool highlands with a climate very different from that of a Canadian summer. Lower temperatures may well account for the East African results.

In New Zealand (Martens *et al.*, 1977) 59 out of 60 isolates of *P. graminis avenae* were virulent for gene *Pg9* and none for gene *Pg8*. In Australia (Luig and Baker, 1973), 13.7% in 1970/1971 and 5.3% in 1971/1972 of the isolates were virulent for gene *Pg9* and none for *Pg8*. In the absence of virulence for gene *Pg8* no convincing analysis is possible, but at least the data show dissociated virulence for genes *Pg8* and *9*. Likewise, only avirulence was recorded for gene *Pg13*, so virulence associations and dissociations for genes *Pg8* and *13* were not available for analysis. (It will be remembered that virulences, not avirulences, determine associations and dissociations.)

Assigning gene *Pg8* to the ABC group and genes *Pg9* and *13* to the XYZ group is purely arbitrary, determined by the toss of a coin. The grouping could just as easily have been reversed, with genes *Pg9* and *13* in the ABC and gene *Pg8* in the XYZ group. Eventually, however, the oat stem rust system will, we believe, be aligned with the wheat stem rust system, presumably when there is information about the quaternary structure of the relevant proteins.

2.21 RESULTS WITH SOME OTHER PATHOGENS

With wheat leaf rust caused by *Puccinia recondita* there is evidence in the data of Samborski (1972a,b, 1974, 1975, 1976a,b, 1978, 1979) that in Canada virulence for genes *Lr17* and *Lr18* was dissociated but that the dissociation was erratic. There was evidence for dissociation in the years 1971 through 1976, but not in the years 1977 and 1978. There is no direct evidence for dissociation in the United States or elsewhere. Indeed, in Nebraska clones virulent on plants with these two genes are common (Karanmal *et al.*, 1978), as they are also in the Pacific Northwest (Milus and Line, 1980). There may be a situation here like that with virulence for wheat stem rust resistance genes *Sr6* and *Sr9d:* Virulence is associated in the United States but dissociated in Canada. We shall have to await evidence from samples taken from mature plants at the end of the season before a realistic analysis is possible.

With stripe (yellow) rust of wheat caused by *Puccinia striiformis* there is evidence of dissociation of virulence matching resistance genes *Yr1* and *Yr4* (Scott *et al.*, 1979). With barley mildew caused by *Erysiphe graminis hordei* there is evidence of dissociation of virulence matching resistance genes *BMR5* and *6* (Scott *et al.*, 1979).

In the data so far available, virulence patterns are clearer in *Puccinia graminis* than in *P. recondita, P. striiformis,* or *Erysiphe graminis*. In part we may possibly ascribe this to the greater abundance of relevant data about *P. graminis*. The clearer pattern could, however, also possibly be due to differences in the nature of the four diseases: stem rust is likely to be sampled later in the season, and therefore after warmer temperatures, than leaf rust, stripe rust, or powdery mildew.

In populations of *Phytophthora infestans* there is evidence for various virulence associations. Among others, virulence for potato blight resistance genes *R5* and *R7* was commonly associated with virulence for gene *R2* in the surveys of Malcolmson (1969) conducted widely over Great Britain and invariably associated in the surveys of Shattock *et al.* (1977) conducted in Wales. Some discussion is necessary, because Shattock *et al.* (1977) assert, in the face of their own evidence, that virulence distributions were random.

Statistical evidence about virulence for genes *R2* and *R7,* taken from the surveys of Shattock *et al.* (1977) for the four years, 1970 through 1973, is given in Table 2.25. The association of virulence is not in doubt. As background, note that no potato variety grown in Britain has the gene *R7* (or *R5*). Virulence is unnecessary (see Section 2.4). A few, not commercially very important, varieties have the gene *R2*. But we can exclude their being directly responsible for the virulence associations, for two reasons. First, of the total (200) isolates of *P. infestans* analyzed by Shattock *et al.* (1977), only one-third (66) was virulent for

TABLE 2.25

The Number of Isolates of *Phytophthora infestans* Virulent and Avirulent for the Potato Blight Resistance Genes *R2* and *R7*[a]

	Virulent for *R7*	Avirulent for *R7*	Total
Virulent for *R2*	48 (15.84)	18 (50.16)	66
Avirulent for *R2*	0 (32.16)	134 (101.84)	134
Total	48	152	200

[a] Data of Shattock *et al.* (1977).

[b] Figures in parentheses are numbers expected if the distribution were random. $\chi^2 = 128.24$, $n = 1$. For $n = 1$ and $P = 0.001$, $\chi^2 = 10.83$. The probability of a random distribution is therefore vanishingly small.

gene *R2;* the other two-thirds were all avirulent not only for *R2* but also for *R5* and *R7*. Second, virulence for gene *R4* (the commonest virulence) was randomly distributed in relation to virulence for *R2*.

Malcolmson's data about virulence for genes *R2*, *R5*, and *R7* confirm those of Shattock *et al.* (1977). This is true even when isolates of *P. infestans* from the potato variety Pentland Dell are excluded, Pentland Dell having the gene *R2*. These associations of virulence for genes *R2*, *R5*, and *R7* are not the only associations to be found in the potato blight data. But the relevant data for potato blight are much less abundant than those for wheat stem rust, and tell the story less well. So, apart from establishing that in populations of *P. infestans* there is indeed a structure of virulence that is not an immediate response to host resistance, we pursue the matter no further.

2.22 SOME CONCLUSIONS

What emerges clearly from the data in this chapter is that virulence in populations of *P. graminis* has a self-imposed structure. Virulence can be divided into two groups matching two groups of resistance genes. Virulence matching resistance genes of the ABC group dissociates from that matching resistance genes of the XYZ group. Within each group, the virulences associate. There is other virulence, apparently neutral, matching resistance genes not shown to belong to either the ABC or the XYZ groups. This neutrality may be real, or it may simply reflect defective survey procedures. Possibly when better surveys are made more genes will be assigned to one or other group.

From the hypothesis stated in Chapter 6 it is a prediction, which can easily be tested by experiment, that increasing temperature increases the dissociation of virulence for resistance genes of the ABC group from virulence for genes of the XYZ group. Suitable experimental tests would be through surveys conducted along the lines suggested in Section 3.6. When surveys are properly linked to temperature and other environmental factors, many of the present obscurities will disappear.

In terms of practical wheat breeding against stem rust, at least two resistance genes are needed, one selected from each group. Selected in this way, both resistance genes are strong in the language of the second gene-for-gene hypothesis; that is, they are more useful than average to the wheat breeder, other things being equal. On the other hand, two resistance genes from the same group, if only two genes were used, would each be ultraweak.

In terms of population genetics, the results throw out models that have been proposed to explain population changes in pathogens on preconceived notions. Basically we shall ultimately have to explain why the association of unnecessary virulence increases the fitness of a clone of *P. graminis* when the virulences match resistance genes belonging to the same group, but decreases fitness when the resistance genes belong to different groups. The answer will probably emerge when the quaternary structure of the relevant proteins becomes known.

Present surveys of *P. graminis* populations are inadequate. This emerges clearly from the data. The Canadian–United States frontier is a political, not ecological, boundary; and the abrupt reversals at this boundary illustrated, e.g., in Table 2.2, show that in one or both of the surveys environmental effects are inadequately sampled. Two sources of inadequacy need comment.

First, the surveys implicitly assume that populations of *P. graminis* are fixed for the season. This is not explicitly stated in the literature of the surveys, but is clearly implied by the omission of information about the dates when the samples were collected. Hundreds or thousands of samples are collected yearly, without dates being assigned. There are not even broad indications about whether collections were made early or late in the season. This notion of fixity is clearly false; and one need only refer again to Table 2.2 as an example to demonstrate the falsity. There is a need both to consider dates (and therefore temperature and perhaps length of day) as an important variable and to extend surveys to the end of the season.

Second, in the surveys of the pathogen the recorded units are physiologic races, which (in our present context) are taxa fixed by avirulence and virulence for particular resistance genes in the host. By making races the survey's units it is implied that genetic change in the pathogen's population is best studied in fixed aggregates of virulence and avirulence. This is in contrast with the concept of gene flow, in which the gene (or allele) is itself the unit. For more than half a century plant pathology has been in the grip of race theory. It is time to probe that theory.

3

Races of Pathogens

3.1 INTRODUCTION

A race is a taxon too low in status to need a latinized name. Variety, cultivar, strain, biotype, type, and breed (of domestic animal) are other terms for these lowly taxa. This would seem a simple statement of an uncomplicated concept. In point of fact the taxonomy of races in plant pathology is totally unlike taxonomy elsewhere in biology; and if we are to understand the tangle in which plant pathology now finds itself, we must begin at the beginning. Again, we shall use *Puccinia graminis* as the main vehicle for discussion.

Stakman *et al*. (1934) divided *P. graminis* into five formae speciales according to the telial host: *tritici, avenae, secalis, agrostitidis,* and *poae*. The forms rate being latinized. *Puccinia graminis tritici* attacks wheat, barley, and some wild grasses; *P. graminis avenae* attacks oats and some wild grasses, but not wheat or barley, etc. Within the form *tritici* there are races that differ in their ability to attack different varieties of wheat. It is at this level that the special difficulties we discuss begin.

To identify races Stakman *et al*. (1962), revising keys going back to 1922 (Stakman and Levine, 1922), used 12 differential host plants which were varieties of *Triticum* spp. The reaction of each of these differential hosts to infection by a race of *P. graminis tritici* was classified in seven types: 0 (immune), 0;, (almost immune), 1 (very resistant), 2 (moderately resistant), X (intermediate or mesothetic), 3 (moderately susceptible), and 4 (very susceptible). This scheme allows the identification of more than 30 million races.

With the discovery of more stem rust resistance genes in wheat this scheme became inadequate; and with the development of wheat lines each with a different resistance gene (*Sr* gene) it became possible to identify races of *P. graminis tritici* in terms of the *Sr* genes they match. At the same time the scheme was simplified by recording reactions as either resistant or susceptible, with relatively seldom the record of an intermediate reaction. But even this simplification, with the omission of intermediates, leaves the potential number of races far beyond what can be coped with in practice. There are at least 30 known *Sr* genes, and races classified as virulent or avirulent on each of them could number 2^{30} or approximately one billion. (The actual number found in a survey is necessarily limited, among other things, by the limited number of isolates feasibly analyzed in surveys. If as many as a thousand isolates were analyzed a year, it would still take a least a million years to identify the potential billion races. The ABC-XYZ groups and resistant wheat varieties would extend the time even further.)

3.2 ADDITIVE AND MULTIPLICATIVE INCREASE

Standard taxonomy, that is, taxonomy as it is normally known in zoology, botany, and mycology, proceeds by addition. As new taxa are found they are added one by one to existing taxa. Discover a new species of *Puccinia,* and you add one to the known number, without necessarily altering the status and definition of any previously known species. Find a new race of mankind in a hitherto unexplored jungle, and you add one to the known races without necessarily upsetting any previous classification. Breed a new variety of rose, and you add one to the known list of rose varieties. The process of discovery is always one of addition, without necessarily any retrospective adjustment of established taxa.

Pathogen races like those of *Puccinia graminis* discussed in the previous chapter increase by multiplication, and the discovery of a new race automatically affects the standing of all previously known races. It acts retrospectively. For simplicity, consider only virulence and avirulence, without intermediates. Suppose the host plant has two resistance genes *R1* and *R2* in independent loci. There would then potentially be four races of the pathogen: race 0 unable to attack plants with either of these resistance genes, race 1 able to attack plants with gene *R1* but not gene *R2,* race 2 able to attack plants with gene *R2* but not gene *R1,* and race 1,2 able to attack plants with both *R1* and *R2*. Now suppose a third resistance gene *R3* is discovered. All isolates of the pathogen previously assigned to the original four races would have to be reassessed. It would have to be determined whether an isolate previously assigned to race 0 belonged to race 0 or race 3; whether one previously assigned to race 1 belonged to race 1 or race 1,3; whether one previously assigned to race 2 belonged to race 2 or race 2,3; and whether one previously assigned to race 1,2 belonged to race 1,2 or race 1,2,3. There would now be potentially eight races; increasing the number of resistance genes from 2 to 3 would increase the number of races from 2^2 to 2^3. The process of discovering new genes would be one of multiplication, always with retrospective adjustment of previous races. When 30 resistance genes are known, as with wheat stem rust, there would be potentially a billion races. The discovery of the thirty-first gene would, by retrospective action, add a further billion: a single gene would add a billion races. This is absurd.

A change of name, to distinguish multiplicative from additive races, would not help. A billion pathotypes would be just as intractable as a billion races. The fault lies in the multiplicative system itself.

3.3 GENES AND TAXA

In true (additive) taxonomy the number of taxa cannot exceed the number of relevant genes or alleles. By relevant genes we mean those that determine the characters used to distinguish taxa.

The number of great taxa, such as phyla, is small compared with the number of relevant genes. As the status of the taxon is reduced, so is the disparity between the number of taxa and the number of relevant genes. Finally, in the lowest taxa such as color variants of some garden flowers, a variety may differ from others by a single relevant allele. But always in additive taxonomy the rule remains that the number of taxa does not exceed the number of relevant genes or alleles.

In plant pathology this rule is adhered to when the races are horizontal, i.e., when the races differ in aggressiveness. For defintions, see Chapter 5. It is in vertical resistance that races are multiplicative, and, uniquely in biology, often far exceed in number the genes or alleles that determine them. The difference derives from the definition of horizontal and vertical resistance. It reflects a fundamental distinction.

3.4 THE INEPTNESS OF FIXED RACES

Taxonomy is founded on associations of characters that the taxonomist considers to be fixed enough to be diagnostic and used for recognition. Fixity of the characters relevant to the taxon is assumed. One assumes that linnaean species and genera are still the same, after centuries, as when Linnaeus first described them. One assumes that the seed one buys from a seedsman remains fixed from year to year for the varietial character one has in mind.

The fixity of the relevant characters of pathologic (physiologic) races is not in doubt. It is a fact. Races of, say, *Puccinia graminis tritici* are defined by the relevant *Sr* genes for which they are virulent or avirulent (or intermediate). A change from virulence to avirulence, or vice versa, inevitably creates a new race. The argument is not about fixity but about fixity's aptness in problems of epidemiology.

Canada gets its inoculum of *Puccinia graminis tritici* from the United States. Yet races of this pathogen found in substantial amounts in the United States disappear at the Canadian border; and races unrecorded in the United States appear in substantial amounts in Canada. Fixity is not in evidence in the population. Table 2.2 records one such substantial change, from association to dissociation of virulences for genes *Sr6* and *Sr9d*. Note not only that there was a change, but also that in order to demonstrate this change it was necessary to dissolve the races and crystallize particular virulences and, especially, a particular change (flow) from association to dissociation of virulences.

Roelfs (1974) compared two populations of *P. graminis tritici,* and his deductions show the danger of emphasizing fixity. Similar wheat nurseries were planted at Beeville, Texas, and Rosemount, Minnesota. An epidemic developed in both nurseries, but there were major differences between south and north. In particular, the race 15 group was abundant in the north but absent in the south. From this Roelfs postulated that there are in the United States two populations of *P. graminis tritici,* and that these two maintain themselves in the following manner. The southern population persists in southern Texas and Mexico throughout the year, with some northward spread in the spring. The northern population overwinters north of the 30th parallel on hard red winter wheats that have resistance to the southern population but are susceptible to the northern population. The danger in Roelfs' argument is that the same argument could be applied to the results given in Tables 2.2 and 2.4, which show sharp differences between the populations in the United States and Canada, to postulate that there are two populations of *P. graminis tritici* in North America, one persisting south of the border and the other persisting in Canada. This postulate is in conflict with all other evidence. The danger of misinterpretation arises from thinking in terms of populations of fixed races. Consider race 15 that Roelfs singles out. This race found in Minnesota but not in Texas, is virulent for genes *Sr7b, 9d, 9e, 10, 11,* and *Tt1,* i.e., for genes of the XYZ group, and avirulent for genes *Sr6, 9a,* and *9b,* i.e., for genes of the ABC group. Focus attention on gene *Sr9e*. Avirulence for it was universal in the Texas nursery; virulence for it was common in the Minnesota nursery, race 15 making up 87% of the isolates obtained there. An important change, therefore, from Texas to Minnesota was from avirulence to virulence for gene *Sr9e*. (The other races, QFB and RKQ, making up the remaining 13% of the isolates in Minnesota were also present in Texas.) Other changes, such as the change from frequent virulence for genes *Sr6* and *9b* in Texas to avirulence in Minnesota, follow from the dissociation of virulences discussed in the previous chapter. To focus attention on gene flow, instead of on changed races, concentrates thought on the essentials of the problem.

3.5 CHANGE AS THE RESULT OF GENE FLOW

Fluidity, the opposite of fixity, is reflected as gene flow. The population of the pathogen is seen not in terms of fixed associations and dissociations of virulences and avirulences but in terms of fluid associations and dissociations. This brings the topic into the mainstream of population genetics.

Consider, by way of introduction, the classic example of gene arrangements in the third chromosome of *Drosophila pseudoobscura*. Dobzhansky (1943), in one of his studies of the genetics of natural populations, determined the frequency of various gene arrangements, among them "Chiricahua" and "Standard." The adaptive value of these arrangements varied from locality to locality and from month to month. In a 4-year study on the San Jacinto Mountains the observed frequency of the Chiricahua arrangement rose month by month from 11.0% in January, when the flies emerged after the winter to 34.6% in June, and then fell again to 10.1% in October. Contrariwise the frequency of the Standard arrangement fell from 57.5% in January to 37.4% in June and then rose again to 58.9% in October. These and other changes in frequency were constantly recurring and completely reversible. There were evidently spatial and temporal adaptations to the environment, and great changes in the frequency of gene arrangements could occur within two sexual generations. Dobzhansky calculated that the selection coefficient s was as high as 0.4.

The change in *Puccinia graminis tritici* from association to dissociation of virulence for genes *Sr6* and *Sr9d* at the Canadian border is far swifter than anything recorded by Dobzhansky in *Drosophila pseudoobscura,* and the selection coefficient against association must be substantially greater than 0.4. This is not surmise, but hard experimental fact recorded in the surveys summarized in Table 2.2. Other swift changes from association to dissociation are recorded in Table 2.4. How are such swift changes possible? The answer seems to lie partly in asexual propagation of the pathogen and partly in the grouping of virulences for the ABC and XYZ genes.

Asexual propagation, by uredospores in the case of *Puccinia graminis,* allows adaptations in one generation to be passed on without dilution to the following generation. Adaptations do not carry the genetic load inevitable with sexual propagation. Asexual propagation, by allowing quick adaptation to changing environment, allows quick change. An erroneous notion pervades the literature.

Asexual propagation, it is said, promotes constancy; sexual propagation promotes variation. This notion is all very well in the context of the clonal propagation of roses or fruit trees; adaptation here is not closely tied to propagation. But in the context of microbiology adaptation to environmental variation is swiftest if propagation is asexual. Sexual recombination is lauded throughout the literature as the source of variation; but sexual recombination reduces adaptation to quick environmental variation and therefore the variation of the organism itself. A change of thinking is needed, with a better assessment of the inherent advantages of asexual propagation. Instead of thinking of a pathogen population waiting for appropriate mutants or recombinants to arise, one should think of a population in which all the variants are present and needing only selection pressures to increase or decrease their frequency (see Section 4.5). In a process of increasing or decreasing frequency under selection pressure asexual will outstrip sexual propagation. Fact is on the side of this argument. In epidemic disease it is on asexual propagation that fungi primarily rely. Even where facilities for sexual reproduction exists, as for *Puccinia graminis* on *Berberis,* the fungus gets on very well without them.

The ABC–XYZ grouping of virulence speeds change through adaptation. There are (on the gene-for-gene hypothesis) at least 30 loci for virulence/avirulence in *P. graminis tritici*. If each locus was independently and differently adapted to the environment, only one clone in a billion could possibly achieve optimal adaptation at any time, and with the environment changing swiftly from time to time even this adaptation could never be close. Consolidation of adaptations, as in ABC and XYZ groups, avoids the difficulty. Indeed it is evident that adaptations must be consolidated if they are to be close in a quickly changing environment, such as occurs when winter turns to summer in North America.

3.6 COMPUTERIZED SURVEYS

The problem is not only to determine gene flow but also to determine what makes the genes flow. Inevitably attention returns to surveys and how to make them fit the existence of a multitude of relevant resistance genes.

In relation to wheat stem rust, for example, one needs to know in the survey the phenotypic avirulence or virulence of the pathogen for each selected *Sr* gene, the genotype of wheat from which the isolate of *P. graminis tritici* was collected, the latitude, longitude, and if relevant, the altitude of the site of collection, the date of collection, the temperature in appropriate detail for, say, 10 days and 60 days previous to collection, and whatever else is thought to be relevant. Data must be collected up to the very end of the growing season. From these data one needs to be able to correlate the frequency of virulence and avirulence, in dissociations and associations of varying size, with changes in date, temperature, or any other independent variable.

For diseases like wheat stem rust against which there is a large number of resistance genes, only a computer can handle the input and retrieval of information on the required scale. This does not necessarily imply that more isolates are needed, although extra data are always welcome. It does imply that the surveys need machinery for better analysis.

Computers can make (multiplicative) races obsolete. The recorded units in a computer are virulences and avirulences which may increase by addition but not by multiplication. It is unnecessary to determine all possible combinations of virulences and avirulences, i.e., it is unnecessary to determine all possible races. Instead we need only extract the combinations that seem relevant, and need not give these combinations an identifying name or number. For example, an obvious project would be to determine how association and dissociation of virulence for genes *Sr6* and *Sr9d* are affected by temperature changes from January to August. Virulence or avirulence for other genes need be brought into this simplified project only in order to determine whether they are relevant genetic background. At all events it would be unnecessary to identify the potential three-fourths of a billion races (if one assumes that there are 30 *Sr* genes) which are virulent for either or both *Sr6* and *Sr9d*.

3.7 DISCUSSION

The first key to the identification of physiologic races of a pathogen was published in 1922. The pathogen was *Puccinia graminis tritici*. Since then generations of plant pathologists have grown up to accept the system of races without question. We question it now, both because it was misconceived and because its usefulness has been overrated.

The misconception arose from the mistaken belief that in using physiologic races plant pathologists were following general practice in biology. This belief has been explicitly stated by Stakman and Harrar (1957). Visible characters (they point out) are used to group plant pathogenic fungi into divisions, subdivisions, classes, orders, families, genera, species, and varieties. It is easy to distinguish larger groups of parasitic fungi, such as smuts and rusts. But there are many kinds of smuts, such as loose smut of wheat, the stinking smuts of wheat, maize smut, and rye smut. There are more than 3000 kinds of plant rusts. The smaller groups, such as species and varieties, are more difficult to distinguish. But the plant pathologist must not only distinguish them; he must also know their pathogenicity for the various sorts of cultivated and wild plants. The species *Puccinia graminis,* for example, can be distinguished by its general appearance and by some characters of its spores. Within this species are several varieties or *formae speciales* differing in the crop plants they attack. The variety *tritici* attacks wheat, barley, and many other grasses. Within the variety *tritici* are races that differ in their capacity to attack different varieties of wheat; and these races can in turn be subdivided into the ultimate subdivisions, the biotypes. In their argument Stakman and Harrar explicitly liken the genera, species, varieties, and races of pathogenic fungi to the genera of crop plants, the species of *Triticum* (the common bread wheat *T. vulgare,* the durum wheats *T. durum,* and other species and subspecies), the thousands of varieties of bread wheats and the hundreds of varieties of durum, and the new lines continually being produced by plant breeders. Herein their error lies. The sequence of subdivision of plants through *Triticum* down to the latest line of bread wheat is always a sequence in additive taxonomy. The sequence of pathogenic fungi through *Puccinia* down to races and biotypes is a sequence in additive taxonomy as far as varieties or *formae speciales* and is thereafter a sequence in multiplicative taxonomy. One cannot prop, as Stakman and Harrar tried to prop, the concept of pathogenic races by appeal to general taxonomy.

The usefulness of the concept of pathogenic races has been overrated through confusing in the results of race surveys the usefulness of the survey with the usefulness of the concept of races. Canada gets most of its inoculum of *Puccinia graminis* from the United States. Yet many of the isolates of *P. graminis tritici* in Canada belong to races rare or unknown in the United States. United States races disappear at the border, and Canadian races arise anew. This comparison is valid even when confined to avirulence and virulence for *Sr* genes common to both surveys. The surveys as such are unquestionably useful. It is unquestionably useful for a wheat breeder in Canada to know that, in 1978, all isolates of *P. graminis tritici* in the survey were virulent for gene *Sr14,* and none virulent for *Sr22, Sr24, Sr26, Sr27, Sr29,* or *Sr30* or for *Sr6* and *Sr9d* in combination. But it is an unwarranted extrapolation to assume that this justifies the grouping of virulence in fixed races. On the contrary, the footnotes of Tables 2.15, 2.16, and 2.17 show how necessary it is to strip races in order to learn what we want to know.

Physiologic races will be in the literature for years to come, if only because the past literature has been recorded in race form. Whatever one's opinion of the concept of physiologic races may be, one is bound by the past, and it would be pedantry for a writer not to refer to literature in the form which through custom is most easily assimilated by the reader. Computers will replace much of race literature, but they have not yet been relevantly used. A changeover in custom is going to be a long process.

As a last point on a different theme, phenotype analysis is all very well, but sooner or later it will have to be supplemented by at least some genetic analysis. In the previous chapter the ABC–XYZ grouping in populations of *Puccinia graminis tritici* was seen to be dictated by virulence rather than avirulence. But avirulence is ambivalent. Virulence is usually recessive and avirulence often heterozygous. The comparisons in the previous chapter, between virulence and avirulence, were therefore probably largely comparisons between double and single doses of the virulence allele, and not between the presence or absence of the allele. For a finer analysis of the ABC–XYZ grouping we shall have to be able to distinguish between heterozygous and homozygous avirulence; and this will necessitate associating surveys with at least a modicum of genetic analysis.

4

The Influence of the Host

4.1 INTRODUCTION

Chapters 2 and 4 are both concerned with the influence of the host plants on the pathogen, but with a different emphasis. In Chapter 2 the influence of the host was not emphasized; the emphasis was put on the interaction of virulence and avirulence in the population of the pathogen, and the genotype of a host plant was seldom mentioned. This does not imply that there was no influence of the host in the background. The virulence structure of the pathogen population must depend both on the host population and on internal selection within the pathogen population itself, and Chapter 2 was concerned primarily with the pathogen. Now, in Chapter 4, the emphasis is changed in order to write more about the influence of the host plants.

The boom-and-bust cycle in plant breeding for resistance has been widely discussed in the literature of the past half century. A crop is susceptible; the plant breeder develops a resistant variety, which introduces a boom period while the resistance lasts; the pathogen adapts itself to the resistant variety by appropriate changes to virulence; resistance is lost and the erstwhile resistant variety becomes susceptible. The literature of examples of evanescent resistance in different crops throughout the world would fill a book.

Resistance to disease caused by biotrophic fungi has been especially vulnerable. But the topic is so well known that it will be unnecessary to mention more than a few case histories, simply to illustrate how changes in the host have brought about changes in the pathogen.

The other side of the picture is more important. Resistance has often been successfully and stably introduced by plant breeders, and the literature of adequately stable resistance would also fill a book. Resistance in the host plants has not brought about matching changes in the pathogen. The position at the moment is that maize, wheat and other grains, sugarcane, beet, and many other field crops are being grown with little or no protection by fungicides except for seed and planting material; and they are being grown successfully because of the conscious and unconscious selection for resistance against disease. Much of this resistance is horizontal and therefore expected to be stable. But some of it is vertical, and there are millions of hectares of crops grown successfully under the protection of vertical resistance.

In vertical resistance it is the story of lack of influence of changes in the host, i.e., it is the story of stable vertical resistance, that has special interest, and much of this chapter is about this. We start however with examples of unstable resistance.

4.2 VIRUS DISEASES

Resistance to virus diseases has been widely used in agriculture, and the resistance against most of these diseases has remained stable. However, there are exceptions, notably tomato mosaic caused by tobacco mosaic virus. Pelham *et al.* (1970) have demonstrated a strong influence of the host on the pathogen in Britain. Prior to 1966 only avirulent strains were found. In 1966 resistant tomato varieties, with the gene *Tm1*, were introduced. By 1967 the frequency of virulent strains in greenhouses growing these resistant varieties had increased to 50%, and by 1968 to 93%.

Tomato spotted wilt is another virus existing as a number of strains under different genetic control (Finlay, 1952). Tomato varieties resistant in the Hawaiian Islands and New Jersey were found to be susceptible in Western Australia. Yet despite this variability there are few records of tomato varieties becoming susceptible where they were previously resistant: this stability is perhaps the result of spotted wilt being essentially exodemic in tomatoes.

4.3 BACTERIAL DISEASES

In bacterial diseases evidence that the introduction of resistant host plants brings about a change toward more frequent virulence in the pathogen has been noted for blight of cotton caused by *Xanthomonas malvacearum*. Bacterial blight is the major disease of cotton in Africa and elsewhere were wind-driven rain or overhead irrigation disperses the pathogen. There are various control measures, among them the use of resistant varieties. Breeding cotton for resistance to *X. malvacearum* has been a major project in Africa for half a century, and is a common project elsewhere. The literature was reviewed by Brinkerhoff (1970). When resistance was conditioned by few genes it was often unstable. Thus in the United States the resistance of cotton varieties with single *B* genes in a susceptible background was matched in a single season by increased virulence in *X. malvacearum*. Indeed, Brinkerhoff (1963) found that inoculum from Uganda was virulent for all the genes *B1, B2, B3, B4, B5, B7, Bin,* and *Bn*. Cross (1963) in East Africa found that the resistant variety Albar 51 and its derivatives were matched by appropriate cultures of the bacterium. Nevertheless, cotton breeders seem to have succeeded in synthesizing high levels of resistance which appears to be stable.

Bacterial blight of rice caused by *Xanthomonas oryzae* is found in most rice-growing areas of Asia and has recently been reported in America. It is a highly variable organism, and breeding for resistance has been undertaken in many countries. There have been reports of the breakdown of resistance (Rao *et al.*, 1971; Reddy and Ou, 1976), but these have been surprisingly few, considering the variability of the pathogen.

A destructive adaptation of bacteria to the host was reported by Crosse (1975) in England. Before 1971 the variety Roundel of sweet cherry was substantially resistant to *Pseudomonas mors-prunorum,* but in 1971 a variant of this bacterium arose which differed pathogenically from other strains and was highly destructive on Roundel.

4.4 FUNGUS DISEASES

The boom-and-bust cycle in fungus disease has been widely publicized. It would be unprofitable to try to summarize the literature. Instead, examples are drawn from only two diseases: potato blight caused by *Phytophthora infestans* and wheat stem rust caused by *Puccinia graminis tritici*.

Experiments were begun on a substantial scale in the 1920s to introduce *R* genes into the potato. The sources of the genes were *Solanum* spp. other than *S. tuberosum* itself, mainly *S. demissum*. In the 1930s the first commercial varieties were released in Germany, but the rapid increase of area under *R*-gene varieties did not take place until the late 1940s and 1950s. Optimism at that time was intense, but was soon destroyed by the failure of the resistance to be maintained as new virulent blight races became prevalent soon after resistant varieties were released. By the end of the 1960s little faith was left in *R* genes, and potato farmers now rely on fungicides to control blight as much as they had ever done before.

Because *S. tuberosum* itself has no *R* genes the history of change is particularly clear. Early in the 1950s the influence of *R* genes on the population of *P. infestans* was still relatively unimportant, and surveys revealed what one supposes to have been the wild-type population. The only *R* genes then available for purposes of survey were *R1, R2, R3,* and *R4,* and results were of two sorts.

For genes *R1, R2,* and *R3* virulence in populations of *P. infestans* was rare in the early 1950s. Frandsen (1956) analyzed isolates from 34 fields of blighted potatoes in northwestern Germany. None of the potato varieties involved had an *R* gene. Of the 34 isolates, only one was virulent for the gene *R1* and none for the genes *R2* and *R3*. In Canada Graham (1955) analyzed 56 isolates from fields without *R* genes. None of the isolates was virulent for the genes *R1, R2,* or *R3*. These surveys of Frandsen and Graham merely confirmed the fact that virulence for the genes *R1, R2,* and *R3* was rare in *P. infestans,* a fact obvious from the high resistance that new cultivars with one or other of these genes had when they were first released.

For the gene *R4* virulence was present from the very start, even though the gene had not been used by potato breeders. Frandsen's 34 isolates were all virulent for it. Clearly, one had to distinguish between genes *R1, R2,* and *R3,* on the one hand, and *R4,* on the other. Because of the abundance of virulence for gene *R4* in wild-type populations of *P. infestans,* the gene was almost useless for potato breeders, and was called weak (Vanderplank, 1975). (See also the second gene-for-gene hypothesis in Section 2.14.) More recently the genes *R5* to *R11* have become available. They have not been used in potato cultivars in Britain, but virulence for them is nevertheless common in British populations of *P. infestans* (Malcolmson, 1969; Shattock *et al.,* 1977). It would seem that they, like the gene *R4,* are weak.

To return to the genes *R1, R2,* and *R3,* virulence for them built up rapidly when the genes became widespread in potato fields. Blight surveys in the 1960s reported virulence to be common, and they were substantiated by reports of blight in unsprayed fields of varieties with *R* genes. Two histories illustrate the process of the accumulation of virulence.

The potato cultivar Kennebec has the gene *R1,* and some details of the history of blight in it were given by Vanderplank (1968). It was released in the United States in 1948. Although it was not the first variety with the gene *R1* to be released in North America, it was the first to become widely grown. At the time of its release virulence for the gene *R1* was so rare that Kennebec was highly resistant; fields of Kennebec either escaped blight altogether, or developed a few lesions late in the season, too few and too late to do much harm. Farmers had no need to use fungicides. In this boom period production of Kennebec increased fast. By 1954, Kennebec and other *R1* cultivars had increased and accounted for

6.3% of the certified seed grown in Maine. With this increase came trouble, and the start of the bust period of the resistance. In 1954, slight to moderate infection was found in widely separated fields of Kennebec as early as the beginning of July (Webb and Bonde, 1956). With the advance of summer a full epidemic of blight developed on Kennebec. At the Aroostook Farm, Presque Isle, Maine, blight was so severe that it caused 90% defoliation in unsprayed plots before September 4 (Stevenson *et al.*, 1955). Kennebec was no longer very resistant, and races able to attack it were evidently abundant. This suggested that Kennebec was already grown widely enough for appropriate inoculum to overwinter in substantial amount. This suggestion was confirmed by Webb and Bonde (1956). They tested 56 isolates from 15 cull piles of diseased tubers thrown out in the spring of 1955. Of the 56 isolates 30 could attack Kennebec. The reason for the downfall of Kennebec's resistance was obvious: Kennebec's increasing popularity among farmers had brought with it an increase of Kennebec-attacking races. Since that time Kennebec has been regarded by farmers as a susceptible variety needing to be protected against blight by fungicides like any other variety.

A potato variety with a more complex resistance system is Pentland Dell. It has the genes *R1*, *R2*, and *R3*. Released as a commercial variety in Britain in 1961 it soon became popular because of its blight resistance. In 1968 it covered the third highest acreage among the maincrop varieties grown in Britain. Some details have been given by Malcolmson (1969). Until 1967 there were a few reports of blight late in the season, but for all practical purposes Pentland Dell was a resistant variety. In 1967 blight appeared in Pentland Dell in Cornwall at the end of July, just as early as in other maincrop varieties in the area. Extensive infection was soon noted in Devon and Cornwall, and within a month reports of crops with 75% of the foliage destroyed were common. By the end of the 1967 season blight had become evident in 14 counties of England and Wales. In the 1968 season blight in Pentland Dell was found even in Scotland. Pentland Dell's boom period was short; substantial planting of the variety in terms of total potato acreage began in 1964, and the boom was over by 1968. Surveys by Malcolmson (1969) showed that virulence for the genes *R1*, *R2*, and *R3* had become widespread in Britain. The bust of Pentland Dell as a blight resister was complete. Not only was the foliage very susceptible for a maincrop variety, but the tubers were quite unusually susceptible. (Pentland Dell is a good example of the vertifolia effect, defined as the loss of horizontal resistance when new varieties are bred under the protection of vertical resistance.)

To turn now to wheat stem rust, the best known example of adaptation of the pathogen to the host was the epidemic in the spring wheat region of North America in the early 1950s. From 1938 to 1950 the population or *P. graminis tritici* in the region was mostly of races 17, 19, 38, and 56, which were avirulent for the widely grown cultivars of bread wheats and durums. In 1950 race 15B, previously infrequent but virulent for the popular wheat cultivars, became common. It comprised 27% of the isolates identified in the United States in 1950, 41% in 1951, 58% in 1952, and 63% in 1953 (Stakman and Harrar, 1957). If one takes race 15B in Canada to be Race C10 (Green, 1971b), the change from race 56 (=C17) to 15B was a change from avirulence to virulence for genes *Sr9a, 9b, 9d, 9e, 11, 13, 17, Tt1,* and *Tt2*. Something more was said about this in Section 2.6.

The effect of releasing new resistant genotypes of wheat on the virulence of populations of *P. graminis tritici* has been well documented in Australia. The variety Eureka with the gene *Sr6* was released in eastern Australia in 1939. Virulence for the gene *Sr6* was unknown or very rare at that time, and because of its resistance to stem rust Eureka immediately became popular. Later, virulence for the gene *Sr6* increased parallel to the increase of Eureka. By 1945 virulence was so prevalent that Eureka became a very susceptible variety, and its popularity declined so sharply that by 1948 it was a minor variety. Details were given by Watson and Luig (1963). The decline of Eureka coincided with the rise of Gabo and other varieties with the resistance gene *Sr11*. Introduced in 1945 these varieties remained free from stem rust for two years, but in 1948 virulence for gene *Sr11* appeared and increased fast (Watson and Luig, 1963). It is relevant to the topic of Chapter 2 that the frequency of virulence for the gene *Sr11*, in the XYZ group, coincided with the infrequency of virulence for the gene *Sr6*, in the ABC group.

The increasing use of wheat stem rust resistance genes in Australia has been accompanied by an increasing range of virulence in *P. graminis tritici*. This is illustrated in Tables 4.1 and 4.2. Table 4.1 is for the northern part of the eastern Australian wheat belt. Rains following harvest are common; they increase the

TABLE 4.1

The Percentage of Isolates of *Puccinia graminis tritici* in Northern New South Wales and Queensland, Arranged according to the Number of Stem Rust Resistance Genes for Which They Were Virulent[a]

Virulent for	Period		
	1954–1958	1959–1963	1964–1968
0 gene	8.6	0.0	0.0
1 gene	36.6	0.9	0.7
2 genes	54.7	46.2	6.6
3 genes	0.1	48.0	73.4
4 genes	0.0	4.9	12.5
5 genes	0.0	0.0	6.7
6 genes	0.0	0.0	0.0

[a] Data of Luig and Watson (1970).

TABLE 4.2

The Percentage of Isolates of *Puccinia graminis tritici* in Southern New South Wales, Arranged according to the Number of Resistance Genes for Which They Were Virulent[a]

Virulent for	Period 1954-1958	1959-1963	1964-1968
0 gene	15.2	0.2	0.0
1 gene	51.7	7.5	1.4
2 genes	33.1	73.9	46.4
3 genes	0.0	16.0	50.3
4 genes	0.0	2.4	1.0
5 genes	0.0	0.0	0.9
6 genes	0.0	0.0	0.0

[a] Data of Luig and Watson (1970).

hazard of stem rust because the fungus can oversummer on self-sown wheat plants and susceptible native grasses. In the period 1964 through 1968 the most popular cultivars in the north all had more than one resistance gene. Table 4.2 is for the southern part of the eastern Australian wheat belt. Stem rust is not so great a hazard as in the north, and in the period 1964 through 1968 several wheat cultivars had no genes for resistance against stem rust. Nevertheless the trend is the same in both tables; increased virulence accompanies breeding new wheat varieties for resistance.

4.5 THE ROLE OF MUTATIONS

Person *et al.* (1976) have summarized evidence for the view that changes in the pathogen population which follow changes in the host population are the result of changes in selection pressures rather than of changes to new mutations. Even in the absence of a sexual stage, potentially virulent single-mutant genotypes would almost certainly be maintained on a continuing basis in large asexually reproducing populations, and double mutants would probably also be maintained. With this summary we agree, with a possible reservation about virulence for wheat stem rust resistance gene *Sr26* which is mentioned later.

Sexual and parasexual processes need not be discussed. Where they occur they must contribute to the total variation; but the variation seems to be adequate even when they are absent. Something has already been said in Section 3.5 about the advantages of asexual variation to the pathogen in epidemics.

Person *et al.* (1976) base their case on a numerical estimate, using powdery mildew of barley caused by *Erysiphe graminis hordei* as an example. To repeat their numbers, they estimate that there would be 10^9 lesions per hectare, each producing 10^4 conidia per day. In 1975 about 3.5×10^6 hectares were sown to barley in the United States. They accepted a mutation rate of 10^{-7}, giving a total of 3.5×10^{12} mutants in the United States per day, and 3.5×10^5 double mutants per day.

A similar estimate was made by Parlevliet and Zadoks (1977) for *Puccinia recondita*. If 1% of the leaf area of wheat is covered by mature uredosori producing 300 spores per mm^2 per day, the number of uredospores produced per hectare per day would be 10^{11}. With a mutation frequency of 10^{-6} for a given locus this would mean that 100,000 mutants are produced per day in a single hectare of wheat.

The message from these figures is clear, even if one concedes, as one must, that the estimates are rough. Mutations are frequent; it is their ability to survive that determines observable changes in the population of the pathogen.

Consider some experimental results with *Phytophthora infestans* in potatoes. Frequent mutation has been demonstrated. Eide *et al.* (1959) used cultures derived from single spores of race (0) which is avirulent for all known *R* genes, and observed frequent mutation to virulence for the genes *R2*, *R3*, and *R4*. Black (1960) recorded mutation from avirulence to virulence for the gene *R1* and, very frequently, *R4*. Graham *et al.* (1961) recorded mutation from avirulence to virulence for the genes *R3* and *R5*, Denward (1970) for the genes *R3* and *R4*, and Shattock (1976) for the gene *R2*. Earlier, Howatt and Grainger (1955) in a greenhouse inoculation experiment found spectacular evidence for the mutability of *P. infestans*. They established potato plants in pots in a special chamber within a greenhouse. The temperature was controlled so that it never exceeded 21°C. Spray nozzles humidified the chamber every hour, adding the moisture equivalent to 18 mm of rainfall per day. Into this chamber they introduced 10,000 seedlings of *Solanum demissum* × *Solanum tuberosum* crosses. Inoculum of race (0) was introduced, and plants without *R* genes became infected. Some weeks after this it was noticed that blight was spreading to plants not previously attacked. Eventually all the seedlings were blighted, and isolations showed that the fungus now belonged to many races, including some virulent for all four resistance genes, *R1*, *R2*, *R3*, and *R4*, which were then known. Mutation followed by selection in a mixed host population under almost ideal environmental conditions for blight allowed a change to occur from race (0) to race (1, 2, 3, 4) within a few weeks. The ideal conditions did not promote the mutation, but they did promote the survival and therefore the identification of the mutants.

The earliest report of mutation from avirulence to virulence seems to be that of Gassner and Straib (1932) in *Puccinia striiformis*. In a single-spore culture and its progeny mutation to virulence was observed at the rate of 1 in 60,000 to 120,000. The mutant form was able to attack many wheat varieties resistant to the original culture. The mutant was found in 34 cultures and was first observed in greenhouse experiments. Later it was found in experimental plots of wheat.

Flor (1958) detected natural mutations from avirulence to virulence in a culture of *Melampsora lini* heterozygous for avirulence. His experiment will be referred to again later.

In an early report Newton and Johnson (1939) observed a mutation from avirulence to virulence in a culture of *Puccinia graminis tritici*. The record is unusual because it is for mutation during the storage of uredospores in a refrigerator. The possibility of contamination was excluded. Watson and Luig (1968) used a mutant of *P. graminis tritici* with gray-brown uredospores as a precaution against unnoticed contamination, and observed stepwise mutation to virulence for the gene *Sr11*. The scale of the experiment was relatively small, so the mutation rate must have been high. More detail about this experiment is given in Section 6.4. Further evidence for frequent mutation is given by McIntosh (1977).

Flor (1958) comments that the number of mutants obtained in experiments suggests that mutation to virulence in rusts may be more frequent than reports indicate. Mutants are difficult to distinguish from contaminants unless special precautions are adopted; and the techniques often used in studies of pathogenic specialization as well as the genetics of host–parasite interaction impede the detection of mutants.

Mutations arise naturally only in populations of the pathogen on susceptible varieties of the host. Mutants to virulence are therefore hidden and escape notice until such time as they or their progeny come into contact with a previously resistant host. It is this which reconciles the relative rarity of mutations to virulence in agriculture with the experimental evidence for relatively frequent mutations in the laboratory. From the pathogen's point of view, most natural mutations to virulence are necessarily wasted. In the susceptible hosts in which they arise mutants to virulence are subject to selection pressures to which their virulence *qua* virulence is irrelevant and no cause for increased fitness.

4.6 STABILIZING SELECTION

Stabilizing selection or homeostasis is the opposite of directional selection or adaptation. Adaptation is a rule of nature. In the context of this chapter, adaptation in the pathogen is the process whereby the pathogen matches with increased virulence any change in the host toward increased resistance. Stabilizing selection is the process that hinders this adaptation.

Crill (1977) and Leonard and Czochor (1980), in reviews, have sought to restrict the use of the term stabilizing selection to quantitative traits. The restriction can be dismissed at once. First, it splits hairs. Stabilizing selection is selection that stabilizes. Second, the restriction is based on the notion that quantitative traits are polygenic, and qualitative traits oligogenic. This notion was demolished by Thompson (1975). Quantitative traits are often oligogenic, and qualitative traits often conditioned by many genes (see Chapter 8).

The subject of the previous section is that mutation from avirulence to virulence is frequent in large populations of the pathogen. Without forces countering the effects of these mutations, virulence for a resistance gene would replace avirulence, and the host would become susceptible. But it is a fact, and the basis of breeding for vertical resistance, that host plants often retain their resistance despite the known occurrence of matching virulent mutants. One concludes that there is selection that operates to reduce the frequency of phenotypic virulence, and this selection is stabilizing selection.

In the context of our discussion, stabilizing selection arises from the adverse genetic load that virulence must carry if it is to remain distinguishable. Consider a host species and a pathogen species evolving together. By mutation or other means the host plants develop resistance. To survive, the pathogen develops matching genes for virulence. If a virulence gene carried no adverse genetic load, the advantage virulence gives to the population of the pathogen would cause the gene to become fixed in the species; and a fixed gene for virulence would be undetectable. Being undetectable, it would be outside the concern of plant pathologists and plant breeders; we would simply recognize the species as a pathogen. Admittedly, there might occasionally be mutations from virulence back to avirulence, but our techniques are not geared to recognizing rare avirulent spores among millions of virulent spores.

If virulence is not to be fixed in the species, it must carry an adverse genetic load that gives avirulence an advantage when virulence is not needed, i.e., when the pathogen population is on host plants without the appropriate resistance genes. If the adverse genetic load in virulence is heavy, the avirulence gene will tend to be fixed in a pathogen population if it is on host plants without corresponding resistance. This is a condition relevant to most plant breeding for resistance. The plant breeder wishing to introduce a new resistance gene exposes plants of the genotype to infection by the existing population of the pathogen, and if the plants are not attacked, i.e., if avirulence seems fixed in the pathogen, he proceeds to develop the genotype as a new cultivar. Unfortunately, mutation from avirulence to virulence can occur, and virulence may soon become evident in the absence of strong stabilizing selection against it. Virulence carrying a heavy adverse genetic load characterizes the corresponding resistance gene as likely to be strong (see Section 2.16); the heavier the load, the stronger is the resistance gene.

If its load is light and fluctuating, virulence is likely to occur frequently in the pathogen population even when the matching resistance gene is absent from the host population. Such resistance genes are weak. The potato blight resistance genes *R5* to *R11* are examples, because virulence occurs although these genes have not been used by potato breeders. So, too, the resistance gene *R4* is weak, because virulence for it was common even before the gene was used by potato breeders.

Nothing that has been said must be taken to imply that the adverse genetic load of virulence is constant for a particular gene. The load can vary, especially in relation to the ABC–XYZ grouping of resistance genes and to environmental conditions, particularly temperature. These were matters discussed in Chapter 2.

The Australian data in Tables 4.1 and 4.2 show the effects of directional selection and stabilizing selection operating oppositely. Directional selection, adapting the pathogen population to the host population, has two effects. Avirulence becomes rare, or rarer, because the avirulent pathogen is not adapted to resistant host varieties. Thus, to use northern New South Wales and Queensland as an example (see Table 4.1), avirulent isolates and isolates virulent for only one resistance gene in the host together accounted for 45.2% of the pathogen population in the period 1954 through 1958, but for only 0.9% in the following period 1959 through 1963. The other effect, following inevitably, is for virulence to become more common. Whereas in the early period 1954 through 1958 only 0.1% of the isolates were virulent for three or more resistance genes, 92.6% of the isolates were virulent for three or more genes ten years later, after wheat breeders had accumulated resistance genes in the popular cultivars. Simultaneously, stabilizing selection was operating oppositely against unnecessary virulence. In the period 1954 through 1958 it limited virulence so that there was no virulence for more than three genes; in the period 1964 through 1968 there was virulence for five genes but no more. The result, clear in both tables, is a frequency pattern with rare extremes and abundant intermediates: much avirulence and much virulence are rare; intermediate and presumable adequate virulence is common.

Flor's (1953, 1956) data on flax rust in North America anticipated the observations on wheat stem rust in Australia. There were the same opposite selection pressures on the pathogen; adaptation to more resistance genes in the host led to a reduction of avirulence, and selection operating against excess virulence kept virulence in the field within limits. Flor analyzed the surveys of *Melampsora lini* in the North Central States from 1931 to 1951. The flax variety Bison has been susceptible to all isolates of *M. lini* collected in North America since 1931 when surveys started. From 1931 to 1940, when Bison was the leading flax variety, 92% of the isolates of *M. lini* were of races 1, 2, and 3. Race 1 has no virulence, i.e., it carries no pairs of recessive virulence genes, and races 2 and 3 each carry one pair. From 1942 to 1947, when the variety Koto was widely grown, 77% of the isolates were of races 1, 2, 3, and 210. Race 210 differs from race 1 only in its ability to attack Koto. From 1948 to 1951, when Koto and Dakota were leading varieties, 68% of the isolates were of races 166, 180, and 210. These races differ from race 1 in attacking either Koto or Dakota or both; they have correspondingly more virulence genes. Throughout the three periods the dominant races were those possessing the least number of virulence genes compatible with survival, i.e., compatible with the ability to attack the widely grown varieties. Throughout the study the percentage of isolates carrying unnecessary genes for virulence, i.e., genes enabling them to attack varieties not grown commercially, decreased.

These results have special importance because of the great scope for variation in *M. lini*. The fungus is eu-autoecious; and because the uredospores do not overwinter in the North Central States, sexual hybridization probably initiates each year's infection. Selection pressure rather than restricted variation curbs unnecessary virulence.

An incidental point about artifacts was raised by Flor (1953). Results in the field could not be reproduced in the greenhouse. In the field, unnecessary virulence was clearly selected against. In glasshouse tests, there was no evidence for such selection. Race 22 of *M. lini,* which posesses at least 20 pairs of recessive virulence genes, apparently sporulated as abundantly on Bison as did race 1, which possesses none. Unnecessary virulence carries an adverse genetic load in the field which it did not carry in the (undescribed) conditions in a greenhouse. Flor tentatively suggested that, avirulence being usually heterozygous, heterosis could be a factor in adaptation to widely ranging temperatures for germination and infection in the field. (This would tally with the hypothesis discussed in Section 6.13.) Be this as it may, the discrepancy between field and glasshouse results should be a caution to those who try to investigate stabilizing selection against unnecessary virulence by experiments in a greenhouse. Artifacts are dangerously misleading and should be shunned.

Flor's gene-for-gene hypothesis is discussed in Chapter 6. In passing, we might note that it was stabilizing selection that gave it to us. Flor observed that when flax had one gene for resistance the common race of flax rust had one for virulence; when flax had two genes for resistance, the common race of flax rust had two for virulence; and so on, with gene matched by gene in number. Had there been no stabilizing selection against unnecessary virulence, i.e., had there been in the common races of rust genes for virulence over and above those needed to match the resistance genes, the clue that Flor followed would not have been there.

To return to *Puccinia graminis,* almost the same pattern as that shown in Tables 4.1 and 4.2 is shown in Tables 4.3 and 4.4, but in a very different way. Virulence and avirulence in *P. graminis tritici* in Canada are recorded in 2 × 2 independence tables, Table 4.3 for the genes *Sr6* and *Sr10* and Table 4.4 for the genes *Sr11* and *Sr15*. The entries in parentheses show what distribution would be expected if virulence and avirulence acted independently. The observed associated virulence is less common than expected, 82 isolates virulent for the genes *Sr6* and *Sr10* compared with 206, and 124 isolates virulent for the genes *Sr11* and *Sr15* compared with 268 expected. The observed associated avirulence

TABLE 4.3

The Number of Isolates of *Puccinia graminis tritici* in Canada Virulent and Avirulent for the Wheat Stem Rust Resistance Genes *Sr6* and *Sr10* in the Years 1970–1976[a]

	Virulent for *Sr10*	Avirulent for *Sr10*	Total
Virulent for *Sr6*	82 (206.17)[b]	150 (25.83)	232
Avirulent for *Sr6*	1746 (1621.83)	79 (203.17)	1825
Total	1828	229	2057

[a] Data of Green (1971a, 1972a,b, 1974, 1975, 1976a,b).
[b] Entries in parentheses are the numbers expected if the distribution were random. $\chi^2 = 757$.

is also less common than expected, 79 isolates avirulent for the genes *Sr6* and *Sr10* compared with 203, and 150 isolates avirulent for the genes *Sr11* and *Sr15* compared with 294 expected. The observed association of virulence for the one gene with avirulence for the other is correspondingly more common than expected. There were 150 isolates virulent for the gene *Sr6* and avirulent for the gene *Sr10* compared with 26; 1746 isolates virulent for the gene *Sr10* and avirulent for the gene *Sr6* compared with 1622; 1579 isolates virulent for the gene *Sr11* and avirulent for the gene *Sr15* compared with 1435; and 199 isolates virulent for the gene *Sr15* and avirulent for the gene *Sr11* compared with 55 expected. These results are statistically significant beyond all doubt, χ^2 values being 757 and 540 for $n = 1$, compared with $\chi^2 = 10.83$ when $P = 0.001$.

TABLE 4.4

The Number or Isolates of *Puccinia graminis tritici* in Canada Virulent and Avirulent for the Wheat Stem Rust Resistance Genes *Sr11* and *Sr15* in the Years 1970–1976[a]

	Virulent for *Sr15*	Avirulent for *Sr15*	Total
Virulent for *Sr11*	124 (268.06)[b]	1579 (1434.94)	1703
Avirulent for *Sr11*	199 (54.94)	150 (294.06)	349
Total	323	1729	2052

[a] Data from the same sources as those of Table 4.3.
[b] Entries in parentheses are the numbers expected if the distribution were random. $\chi^2 = 540$.

This pattern of lower frequency of the extremes of associated virulence or associated avirulence and higher frequency of the intermediates of virulence associated with avirulence is present in the extreme in Table 2.18; all 790 isolates from Mexico to Canada were either virulent for the gene *Sr9e* and avirulent for the gene *Sr15* or virulent for the gene *Sr15* and avirulent for the *Sr9e,* and none was virulent for both genes or avirulent for both genes. Many other examples can be seen in Tables 2.15, 2.16, and 2.17, provided always that one of the resistance genes belongs to the ABC group and the other to the XYZ group.

It was clear from the evidence in Chapter 2 that the pattern is strongly subject to environmental effects. What is needed are tables like Table 4.3 and 4.4 constructed both from data obtained early in the season, soon after disease becomes evident, and from data obtained late in the season, just before the crop ripens. The tables are likely to show marked differences.

4.7 ASSOCIATED VIRULENCE AND DESTABILIZING SELECTION

The pattern of results in the preceding section, with dissociated virulence and stabilizing selection, could have been expected a priori; but there was nothing to suggest a pattern of associated virulence and destabilizing selection. Although the evidence for associated virulence is clear and beyond dispute, the implications of the evidence for destabilizing selection are still far from being properly understood.

Evidence for the association of virulence, provided that it was for resistance genes all of the ABC group or all of the XYZ group, was given at length in Chapter 2. Virulence for one resistance gene has a selective advantage if it is associated with virulence for another gene within the same group. The genetic load that virulence carries can change from adverse to beneficial according to the company the virulence keeps.

Table 4.5 is constructed like Tables 4.3 and 4.4, and uses virulence and avirulence for two of the resistance genes, *Sr10* and *Sr11,* that appeared in the earlier tables. (Similar results to those in Table 4.5 would have been obtained if the other two resistance genes, *Sr6* and *Sr15,* had been chosen instead.) The pattern in Table 4.5 is the reverse of that in Tables 4.3 and 4.4. The extremes of associated virulence or of associated avirulence are now more frequent than would be expected on an assumption of independence, and the intermediates, with virulence and avirulence associated, are less common.

The pattern of association is especially marked in examples of virulence in *P. graminis tritici* for the resistance genes *Sr9e* and *SrTmp*. Table 2.18 shows that from Mexico to Canada the extremes of associated virulence or of associated avirulence prevailed almost to the total exclusion of intermediates of virulence

TABLE 4.5

The Number of Isolates of *Puccinia graminis tritici* in Canada Virulent and Avirulent for the Wheat Stem Rust Resistance Genes *Sr10* and *Sr11* in the Years 1970–1976[a]

	Virulent for *Sr11*	Avirulent for *Sr11*	Total
Virulent for *Sr10*	1681 (1527.95)[b]	159 (312.05)	1840
Avirulent for *Sr10*	23 (176.05)	189 (35.95)	212
Total	1704	348	2052

[a] Data from the same sources as those of Table 4.3.
[b] Entries in parentheses are the numbers expected if the distribution were random. $\chi^2 = 875$.

and avirulence associated. Tables 2.15 and 2.17 substantiate the prevalence of tight associations. In Canada virulence for the genes *Sr9d* and *Sr9e* is almost invariably associated (see Table 2.15). So too in the spring-wheat region of North America the prevalence of the race 15 complex of *P. graminis tritici* brings about the association of virulence for genes of the XYZ group.

Associated virulence also occurs in Canadian populations of *Puccinia recondita* for the wheat leaf rust resistance genes *Lr3ka* and *LrT* (Samborski, 1979).

For *Phytophthora infestans* Table 2.25, constructed like Table 4.5, brings out the association of virulence for potato blight resistance genes *R2* and *R7*.

Superraces of the pathogen with combined virulence might, of course, be expected as an adaptation to host plants with a combination of resistance genes. But we can exclude this explanation here. Some of the genes mentioned, like *Sr9e* and *R7*, have not been used in cultivars; and in any case the explanation could not cover the associated avirulence that so often accompanies associated virulence. There may be, and probably is, an effect of the host plants in the background, as will be suggested in the next section; but the prime cause must be sought within the pathogen itself.

After the discovery of associated virulence arising from forces within the pathogen itself we must face the consequences of association. One of them is that associated virulence could destabilize the resistance of cultivars just as dissociated virulence helps to stabilize it.

4.8 AN ILLUSTRATIVE SUGGESTION

In the early 1950s race 15B of *Puccinia graminis tritici* suddenly exploded in North America and destroyed vast areas of spring wheat and durums which had been resistant during the 1940s. There has been much speculation about why this happened, because there had been no sudden change of spring-wheat and durum cultivars which could explain the explosion.

Here is another suggestion. The spring wheat and durums were destroyed by stem rust because the resistance gene *SrTmp* was introduced into winter wheat. The gene *SrTmp* is now common in winter wheat; the stem rust inoculum for spring wheat and durums comes from rusted winter wheat; and virulence for the gene *SrTmp* brings by association the virulence found in the 15B complex of stem rust races. The chain is complete.

Details of events of 30 years ago are beyond recall, and it would be futile now to debate them. This is not the intention. The illustration is meant to show how far-reaching, in the literal sense of the term, the effects could be of any tampering with established virulence structure. It also shows that we are probably only at the very beginning of the surprises we can expect from the discovery that virulence at one locus can promote virulence at another.

4.9 EPISTATIC INTERACTION AND MATHEMATICAL MODELS

In the structure of *Puccinia graminis* populations epistatic (nonallelic) interaction overshadows all else, and determines the genetic load carried by virulence. Consider virulence for the gene *Sr9e*. Association with virulence for the gene *Sr6* carries a heavy positive (adverse) genetic load, and association with virulence for the gene *Sr5* no load or only a light load not yet determined. Moreover, the epistatic interactions are greatly affected by the environment, as shown by differences between the Canadian and United States surveys of virulence for the genes *Sr6* and *Sr9d*.

Recently, population genetics has been used in debates about host/pathogen systems. One debate (Leonard, 1977; Sedcole, 1978; Leonard and Czochor, 1978; Fleming, 1980) has been about the dynamics of gene-for-gene relationships. The other debate (Groth, 1976, 1978; Barrett and Wolfe, 1978) has been about whether simple races of the pathogen with few genes for virulence, intermediate races, or complex races with many genes for virulence will predominate in pathogen populations attacking multilines. Models have been constructed, none of which takes into account the facts recorded in the previous paragraph. None of them predicts what is known about the structure of populations of *P. graminis,* and models that do not predict fail in their primary purpose. Doubtlessly, appropriate adjustment of the models could now be introduced, but retrospective adjustment is likely to permit prediction only of what is already known.

A feature of all the models that have been suggested is their lack of robustness. The conclusions to be drawn from the models are very sensitive to the assumptions built into the models. Small changes in the pathogen's reproduction rate or in the assumed genetic feedback between host and pathogen can reverse the conclusions from the model. This seems to be far from the fact of great robustness of the structure shown by *P. graminis* populations in a given environment. One need only remember the dissociation year after year of virulence for the genes *Sr6* and *Sr9d* in Canada. Robust models may perhaps not be worth constructing, because when we know the cause of the robustness the need for models will no longer exist.

4.10 VARIABLE MUTATION RATES

There is evidence that the rate of mutation from avirulence to virulence, as detected by its expression in the phenotype, varies greatly from locus to locus.

Flor (1958) used an F_1 hybrid of *Melampsora lini* heterozygous for (dominant) avirulence at four loci, and found significant differences in mutation rates. At one locus two natural mutants were found in 200,000 uredospores; at another, one natural mutant in 600,000 uredospores; and at the other two, no mutants in 300,000 and 900,000 uredospores, respectively.

Luig (1979) studied mutation in *Puccinia graminis tritici* in natural conditions and also with a mutagen (ethyl methanesulfonate). At the one extreme, avirulence for the genes *Sr5, Sr15, Sr21,* and *Sr9e* was found to have a very high spontaneous mutation rate to virulence, as well as a high rate after treatment with the mutagen. At the other extreme is the stability of avirulence for the gene *Sr26* derived from *Agropyron elongatum*. This gene is in the Australian wheat cultivar Eagle released in 1971, and in the cultivars Kite and Jabiru. The popularity of these cultivars increased swiftly, and by 1974 Eagle covered 530,000 and by 1975 655,000 hectares in Australia. All efforts by agronomists to find stem-rusted plants of these cultivars failed. Also, because it was considered unwise to rely on monogenic resistance, a great effort was made to find a virulent mutant that could be used experimentally by wheat breeders in order to enable them to incorporate other resistance genes. Mutagens were used, all to no avail. Whatever the end of the story may be, the evidence will stand that mutation from avirulence to virulence for the gene *Sr26* was rare.

Luig suggested that *P. graminis tritici* possesses intrinsically different genes for avirulence which he grouped into five classes.

1. Genes like those conditioning avirulence for the genes *Sr5, Sr9e,* and *Sr21,* which mutate readily to virulence in the field and after being subjected to mutagen treatment.
2. Genes like that conditioning avirulence for the gene *SrTt1,* which have a substantial spontaneous mutation rate which was not increased by treatment with ethyl methanesulfamate.

3. Genes like those conditioning avirulence for the genes *Sr6* and *Sr30,* which have a low mutation rate both under natural conditions and after being treated with a mutagen.

4. Genes like those conditioning avirulence for the genes *Sr13, Sr24* (from *Agropyron elongatum*), and *Sr27* (from rye), which rarely mutate to virulence.

5. The gene that conditions avirulence for the gene *Sr26,* which has not been known to mutate to virulence.

Section 6.13.7 suggests a mechanism for wide differences in phenotypic mutation rates.

Luig's fifth category, represented by avirulence for the gene *Sr26,* needs further comment. One could suggest chemical reasons why in certain circumstances a pathogen could not develop an allele for phenotypic virulence. But even if the reasons were valid, the pathogen would still have scope for maneuver. It could delete the avirulence allele by deleting the whole locus. Flor (1960) demonstrated such a deletion; avirulence mutated to virulence in *Melampsora lini* as a result of a chromosomal deletion. It is of course an open question whether the pathogen would gain by this maneuver in the long run. Whatever fitness the pathogen gained by virulence might be balanced by a loss of fitness in feeding, because it seems likely that a locus for avirulence/virulence is also a locus concerned with essential parasitism (Vanderplank, 1975).

4.11 GENETIC, PHENOTYPIC, AND EPIDEMIOLOGICAL MUTATION

Mutations from avirulence to virulence of these three sorts occur in decreasing order of frequency.

Genetic mutation is true mutation. It is a change in the sequence of the base pairs, adenine–thymine and guanine–cytosine. In obligate parasites genetic mutation occurs while the parasite is in the host. It can also occur during dispersion, perhaps especially in spores in the upper atmosphere exposed to high radiation. In nonobligate parasites it can also occur in the saprophytic phase. In any event, genetic mutation from avirulence to virulence is a phenomenon of the parasite alone, the host contributing only in so far as it affects the parasite's environment, e.g., by regulating light and shade.

Phenotypic mutation from avirulence to virulence occurs in a parasite–host partnership. To consider only eukaryotic parasites, some product of the virulence gene and some product of the resistant host must recognize one another, with the result that mutation to virulence means a change in the parasite–host relationship evident even in a single plant. Genetic mutations are lost if they bring about no change in phenotypic parasite–host relations. The mutations discussed in the previous section were phenotypic mutations.

Epidemiological mutations from avirulence to virulence are changes in the parasite population as it occurs on the host population. To take an example from the previous section, phenotypic mutation occurs readily from avirulence to virulence for the genes *Sr9e* and *Sr21*; but Luig (1979) observed that these mutations arise frequently in the Australian rust flora only to die out in a season or two without the virulent form persisting. In the particular conditions observed by Luig and in the particular years of his observations, phenotypic mutation from avirulence to virulence for the genes *Sr9e* and *Sr21* was not followed by epidemiological mutation; in relation to virulence for the genes *Sr9e* and *Sr21* the Australian population of *P. graminis tritici* remained for all practical purposes unmutated. In general it would seem that epidemiological mutation depends more on selection than on genetic or phenotypic mutation. Thus, virulence for the gene *Sr30* is very rare in isolates of *P. graminis tritici* in North America but almost universal in eastern Australia (Luig, 1979), where there was an epidemiological mutation from rarity in the early 1960s to abundance now.

Of the three sorts of mutation, epidemiological mutation is the only form in which the frequency of mutation per unit of pathogen, e.g., per million spores, depends of the area of cultivation of the host. One can picture the pathogen population in a state of recurrent genetic and phenotypic mutation, with recurrent and local flares of virulence in different loci. Some of the flares catch on to become the conflagration we call epidemics. Most of the flares simply die out. This statement is essentially an adaptation of Fisher's (1930) postulate that mutations are recurrent, and any mutations now existing must have occurred many times before in the history of the species. To survive, for the flare to become a conflagration, the pathogen must increase, with progeny exceeding parents in number year by year. Its best chance is when cultivation is extensive.

Genetic mutation and phenotypic mutation from avirulence to virulence are essentially phenomena to be looked for when the pathogen is attacking a susceptible crop, because it is only in pathogens on susceptible crops that avirulence can exist. Because the crops are susceptible, the mutations are not apparent and are easily overlooked. Epidemiological mutation involves the pathogen on resistant varieties, because the mutation would have no epidemiological significance unless it allowed previously resistant varieties to be attacked. (In the foregoing sentences, susceptibility and resistance are defined in relation to the avirulence allele subject to mutation.) For genetic and phenotypic mutants to become epidemiological mutants they must survive first of all in the susceptible crop where they originated; then they must reach a resistant crop, e.g., by windblown spores; and in the resistant crop they must survive from year to year, with all the hazards of survival in the off-season if the crop plants are annuals. The chance of reaching a resistant crop and of surviving from year to year increases with the area under cultivation. This is the host effect in breeding for resistance to disease. A successful breeding program is one that introduces a new resistant cultivar over a large area, giving the pathogen great scope for survival, and does so without incurring epidemiological mutation.

There have been many successful programs. It would be fitting to conclude by mentioning one of them, to illustrate the achievement from great and extended research by plant pathologists and geneticists. Green and Campbell (1979) list ten wheat cultivars licensed in Canada during the past 30 years, and describe their performance in resisting stem rust. Only Canthatch, licensed in 1959, and Pitic 62, licensed in 1969, have become susceptible. Of the rest, seven became cultivated over areas exceeding a million hectares a year in their heyday, without becoming susceptible to stem rust. They are Selkirk licensed in 1953, Pembina licensed in 1959, Manitou licensed in 1965, Neepawa licensed in 1969, Napayo and Glenlea licensed in 1972, and Sinton licensed in 1975. Among these cultivars there have been changes in popularity, but not from any failure to resist stem rust. The example of the Canadian wheat varieties has been quoted, because it illustrates the success achieved with vertical resistance, which from its very nature is more easily neutralized by epidemiological mutation. To quote examples of stability in horizontal resistance would serve little purpose, because horizontal resistance has a built-in safe-guard against being neutralized by epidemiological mutation, however great the rates of genetic and phenotypic mutation might be.

5

Host and Pathogen in a Two-Variable System

5.1 INTRODUCTION

Host-pathogen interactions, the topic of this book, involve the variation of host and pathogen as the foundation of all else. This chapter is intended to strip away what is irrelevant to the very simplest concept of host and pathogen variation. That is, host and pathogen are considered as the only variables, everything else (temperature, crop rotation, etc.) being assumed to be constant.

In such a two-variable system, resistance of the host to disease must exist in two, and no more than two, forms. The main effect of variation of the host, using the term main effect strictly in its biometric sense, determines horizontal resistance. The first-order interaction, i.e., the differential effect, determines vertical resistance. These are definitions, pure and simple, illustrated biometrically. There is no either/or situation. Resistance in any one host plant may be a mixture of horizontal and vertical, in any proportion (including zero and 100%), just as in the analysis of variance main effects and interactions may occur together, in any proportion. To summarize, in the two-variable system, horizontal and vertical resistance, together or separately, describe every possible form of resistance.

So also in a two-variable system, pathogenicity must occur in two, and no more than two, forms. The main effect of variation in the pathogen, again using the term main effect strictly in its biometric sense, determines aggressiveness, and the first-order interaction, i.e., the differential effect, virulence. These are definitions. There is no either/or situation, because main effects and interactions coexist; and pathogenicity may exist as a mixture of aggressiveness and virulence, in any proportion. To summarize, in a two-variable system aggressiveness and virulence, or their mixture, describe every possible form of pathogenicity.

To put the matter in another way for clarity, horizontal resistance and vertical resistance, and aggressiveness and virulence are precisely defined terms derived to cover all possibilities within a two-variable system of host and pathogen.

The definitions of horizontal and vertical resistance started, and are still, purely mathematical in concept. They have epidemiological and biochemical implication; but these are implications only, and do not enter the definitions or this chapter. To illustrate the concepts of horizontal and vertical resistance it is necessary at times to involve a genetic background, but the definitions are not genetic.

Since the concept of horizontal and vertical resistance was published in 1963, much has been written, sometimes straying from the point. It seems appropriate now to restate briefly the basic concept, and to do it within the framework of two variables.

5.2 THE GEOMETRIC ILLUSTRATION

Vanderplank (1963) used a geometric illustration for the two possible sorts of resistance a host plant can have in a two-variable system. Horizontal resistance is expressed uniformly against races of the pathogen, and vertical resistance is expressed differentially.

Figure 5.1 shows the behavior of two potato varieties, Kennebec and Maritta, infected with *Phytophthora infestans*. Both these varieties have the resistance gene *R1*. This gene confers resistance to races (0), (2), (3), (4), (2, 3), (2, 4), (3, 4), and (2, 3, 4). It does not confer resistance to races (1), (1, 2), (1, 3), (1, 4), (1, 2, 3), (1, 2, 4), (1, 3, 4), and (1, 2, 3, 4). The resistance conferred by the gene *R1* is expressed differentially against the races, and is vertical.

Against the races to which the gene *R1* does not confer resistance, Kennebec and Maritta behave differently. Kennebec succumbs faster to blight than Maritta. Grown side by side, Kennebec is blighted brown while Maritta is still mainly green. Maritta has more "field resistance," to use a common term; Maritta has more horizontal resistance, to use our term. This is illustrated in Fig. 5.1. Where there is no vertical resistance against a race, the horizontal resistance is shown higher for Maritta than for Kennebec.

The resistance of Maritta depends on the composition of the spore shower of

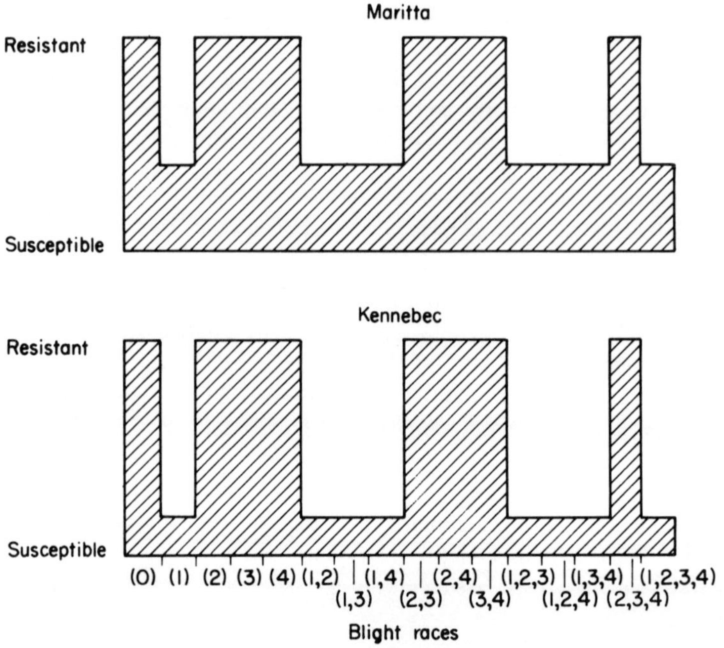

Fig. 5.1. Diagram of the resistance to blight of the foliage of two potato varieties, Kennebec and Maritta, both with the resistance gene *R1*. The resistance is shown shaded to 16 races of blight. To races (0), (2), (3), (4), (2, 3), (2, 4), (3, 4) and (2, 3, 4) the resistance of both varieties is vertical. To races (1), (1, 2), (1, 3), (1, 4), (1, 2, 3), (1, 2, 4), (1, 3, 4) and (1, 2, 3, 4) resistance is horizontal, small in Kennebec and greater in Maritta. From Vanderplank (1963, p. 175; 1968, p. 9).

P. infestans. As almost every survey shows, populations of *P. infestans* are mixtures of races. Against a mixture of races (0) and (1), or against any mixture of races with and without 1 in their designation, Maritta must simultaneously exhibit both horizontal and vertical resistance: horizontal resistance to the race components with 1 and vertical resistance to those without 1 in their designation. Resistance in one and the same Maritta plant at one and the same time would be both horizontal and vertical. By every conceivable agricultural or laboratory test, using *P. infestans* populations from almost every country of the world where potato blight is a menace, Maritta would be resistant both horizontally (by comparison with Kennebec) and vertically, irrespective of whether single plants or whole fields were tested.

Figure 5.1, taken unchanged from the original 1963 publication, shows that there is no either/or condition in the definitions of horizontal and vertical resistance. Both forms of resistance not only can but also must coexist when, as is almost universally the case, the pathogen comes as an appropriate mixture of races. Yet Nelson (1978) founded a long review of horizontal resistance by writing that "no variety can exhibit both HR and VR, by Vanderplank's definitions," and, again, that "Vanderplank's own definitions decree that a plant genotype will express either HR or VR; it cannot express both, by definition." This foundation makes nonsense of the review. Those who prefer to think biometrically will note that Nelson's assertion that horizontal and vertical resistance cannot coexist is an assertion that main effects and interactions cannot coexist; and no further comment on Nelson's review is needed.

5.3 ILLUSTRATION BY ANALYSIS OF VARIANCE

Consider for illustration some data of Paxman (1963) on *P. infestans* in potato tubers. Using potato varieties without *R* genes, Paxman set out to determine whether isolates of *P. infestans* would become specially adapted to a variety if they were grown continously on it. He obtained an isolate 30RS from a naturally infected tuber of the variety Red Skin, another isolate 31KP from a naturally infected tuber of the variety Kerr's Pink, and an isolate 32KE from a naturally infected tuber of the variety King Edward. These he subcultured, each on its original variety, i.e., he subcultured isolate 30RS on tubers of Red Skin, isolate 31KP on tubers of Kerr's Pink, and isolate 32 KE on tubers of King Edward. By the time he came to make his tests there had been 90 cycles of subculturing; for example, isolate 30RS had been on Red Skin for 90 cycles plus the unknown number of cycles the fungus had been on Red Skin in the field before it was isolated. In his tests he used as his criterion the rate of spread of mycelium in tuber tissue. He measured the rate of spread of each of his three isolates not only in the variety of origin, e.g., 30RS in Red Skin, but also in the other two varieties, e.g., 30RS in Kerr's Pink and King Edward as well. In addition, he tested a fourth isolate (of unspecified origin, cultured on the variety Majestic) in the three test varieties; and he did two separate tests. The analysis of variance is given in Table 5.1.

TABLE 5.1

Combined Analysis of Variance in Two Tests of Four Isolates of *Phytophthora infestans* in Three Potato Varieties[a]

	Degrees of freedom	Mean square	*P*
Varieties	2	919.0	<0.001
Isolates	3	563.7	<0.001
Varieties × isolates	6	23.5	
Error	132	19.6	

[a] Data of Paxman (1963).

TABLE 5.2

Analysis of Variance in Relation to Definitions of Horizontal Resistance and Vertical Resistance in the Host, and Aggressiveness and Virulence in the Pathogen in a Two-Variable System

Main effect of host varieties	Horizontal resistance
Main effect of pathogen isolates	Aggressiveness
Interaction, varieties × isolates	Vertical resistance, and virulence

There was a highly significant main effect of varieties, i.e., there was a highly significant difference in horizontal resistance between varieties, with two degrees of freedom. There was a highly significant main effect of isolates, i.e., there was a highly significant difference in aggressiveness between isolates, with three degrees of freedom. But the interaction, varieties × isolates, with six degrees of freedom was insignificant. That is, with the particular varieties and isolates he used, and with his particular measurements of mycelial spread, there was no evidence for vertical resistance or virulence. (It will be remembered that the varieties Paxman used had no *R* genes.)

Also using the technique of the analysis of variance Caten (1970) obtained results very similar to those of Paxman. In tubers Caten measured the surface growth of the fungus in nine days along the tuber from the point of inoculation. There were highly significant main effects of cultivars (i.e., in horizontal resistance) and of isolates (i.e., in aggressiveness); but the interaction cultivars × isolates was insignificant. There was no evidence for vertical resistance and virulence. In leaves he measured the latent period, from inoculation to the first appearance of sporangia. Again there were highly significant main effects of cultivars and of isolates; but the interaction, cultivars × isolates, was insignificant in two of the three tests, and barely significant in the third.

Table 5.2 generalizes and defines the two sorts of resistance and the two sorts of pathogenicity in terms of an analysis of variance, where an analysis of variance is appropriate. Observe that no categories are possible in the two-variable system other than main effects (horizontal resistance and aggressiveness) and first-order interactions (vertical resistance and virulence).

5.4 LIMITATIONS OF THE ANALYSIS OF VARIANCE TECHNIQUE

In the analysis of variance discussed in the previous section two conditions (among others) must be met. First, the unaccountable variation or error in different measurements must have a normal law of error; that is, the frequencies of errors of different sizes must be normally distributed. Second, the relative sizes of the errors in different measurements must be unrelated to any factor in the experiment; the error must be homogeneous.

Johnson and Taylor (1976) and Parlevliet and Zadoks (1977) have used hypothetical sets of data illustrating horizontal resistance, and analyzed them to demonstrate an interaction in conflict with the definitions in Table 5.2. Their sets of data depart far from the conditions (of normal distribution and homogeneous error) laid down in the previous paragraph; and they demonstrated no more than that one should not try to apply the technique of the analysis of variance to data to which the technique is inapplicable.

When resistance and pathogenicity are continuously distributed, appropriate transformations may make the data fit for the analysis of variance, even if they were not originally fit. Johnson and Taylor (1976) showed that for their hypothetical data a logarithmic transformation would do. Tests for whether any particular set of data, with or without transformation, is suitable for an analysis of variance is best left to professional biometricians.

5.5 DEGREES OF FREEDOM AS A LIMITATION

Although the technique known as the analysis of variance cannot always be applied, other methods are often available to detect interactions. One condition is however universal. There must be enough degrees of freedom. Unless there are at least two genotypes for resistance and two for pathogenicity, there will be no degrees of freedom for interactions.

Type A resistance in cabbage to *Fusarium oxysporum* f. sp. *conglutinans* (Walker, 1930) is conditioned by a single gene, and no virulent races of the pathogen are known. There are no degrees of freedom and therefore no possibility of deciding whether the resistance is horizontal or vertical; indeed, there is no need for such a decision. Stable monogenic resistance of this sort is not rare; other well-known examples are resistance in sorghum to *Periconia circinata* and in oats to *Helminthosporium victoriae*.

Consider two examples in which there are degrees of freedom for resistance but none for pathogenicity and therefore none for interaction. Against *Helminthosporium turcicum* Hooker (1963) found a resistance gene *Ht1* in maize. Hooker and Tsung's (1980) survey of 27 inbred lines of maize in the United States show that the gene is now widely used. The gene continues to condition resistance against *H. turcicum,* only two instances of virulence being known. Worldwide tests revealed virulence in isolates from Hawaii, and recently Turner and Johnson (1980) obtained a virulent isolate near Brook, Indiana. Against the Hawaiian isolate a gene *Ht2* confers resistance (Lim *et al.*, 1974). The same gene probably also confers resistance agaainst the Indiana isolate, because maize inbred NN14B which is a source of the gene *Ht2* (Hooker, 1977) is resistant to this isolate. The gene *Ht2* also confers resistance to the common race of *H. turcicum*.

Against the common race of *Fusarium oxysporum* f. sp. *lycopersici* a resistance gene *I* confers resistance in tomato. The common race is called race 1 by Cirulli and Alexander (1966) and race 0 by Gabe (1975). A virulent race called race 2 by Cirulli and Alexander and race 1 by Gabe is known. Against it another resistance gene *I-2* confers resistance. This gene also confers resistance aginst the common race.

Suppose that another race of *H turcicum* was now found that was pathogenic for maize with the gene *Ht2*. There would be two possibilities. If this new race was not pathogenic for maize with the gene *Ht1*, there would be a differential interaction, giving evidence for vertical resistance. But suppose that the new race was pathogenic also for maize with the gene *Ht1*, and suppose that genetic analysis of *H. turcicum* showed this pathogenicity to be monogenic, there would be evidence for a quantitative effect and horizontal resistance.

Much the same argument holds if another race of *F. oxysporum* f. sp. *lycopersici* were found that was pathogenic for the gene *I-2*. There would again be two possibilities. If this new race were not pathogenic for tomatoes with the gene *I*, there would be evidence for vertical resistance. If it were pathogenic for the gene *I* as well as *I-2*, there could be no finality because in the absence of a sexual phase genetic analysis of *F. oxysporum* is impossible; but if many isolates of different origin were found with pathogenicy for the genes *I* and *I-2* combined and never for gene *I-2* alone, one could reasonably suspect that there was a case for horizontal resistance.

To clarify the situation further, Table 5.3 has been constructed. It is supposed that there are three genes for resistance, *R1*, *R2*, and *R3*, and three for pathogenicity, *p1*, *p2* and *p3*. Here *p* stands for virulence if the resistance is vertical, and for aggressiveness if the resistance is horizontal. In the top part A of the table the arrangement is qualitative, i.e., there is an interaction clearly indicating vertical resistance. In the bottom part B the arrangement is quantitative. Resistance is horizontal, as shown by the ranking with resistance increasing column by column from left to right, and pathogenicity increasing row by row from top to bottom. In part B horizontal resistance is taken to be quantitative (i.e., with the genes differing in strength) rather than additive (i.e., with the effect of gene numbers), because this is indicated by comparing genes *Ht1* and *Ht2*, or *I* and *I-2*.

Into Table 5.3 are fitted the data, shown in parentheses, known for maize helminthosporium blight and tomato fusarium wilt. The data fit equally well into either the top or bottom part of the table. In other words, there is nothing which can decide whether resistance in these two diseases is vertical or horizontal; it could well be either. A decision is impossible because there are no degrees of

TABLE 5.3

The Reaction of Plants When Three Resistance Genes *R1, R2,* and *R3* and Three Pathogenicity Genes *p1, p2,* and *p3* Are Arranged Qualitatively (A) and Quantitatively (B)

	Resistance genes		
Pathogen	*R1*	*R2*	*R3*
A. Qualitative arrangement			
Avirulent	(Resistant)[a]	(Resistant)	Resistant
p1	(Susceptible)	(Resistant)	Resistant
p2	Resistant	Susceptible	Resistant
p3	Resistant	Resistant	Susceptible
B. Quantitative arrangement			
Avirulent	(Resistant)[a]	(Resistant)	Resistant
p1	(Susceptible)	(Resistant)	Resistant
p2	Susceptible	Susceptible	Resistant
p3	Susceptible	Susceptible	Susceptible

[a] The reactions in parentheses are known for *Helminthosporium turcicum* in maize and *Fusarium oxysporum* f. sp. *lycopersici* in tomato. In maize genes *R1* and *R2* stand for genes *Ht1* and *Ht2,* and in tomato for genes *I* and *I-2*. In *H. turcicum* the avirulent race is the common race; and *p1* stands for the Hawaiian and Indiana races. In *F. oxysporum* the avirulent race is race 1 *sensu* Cirulli and Alexander or race 0 *sensu* Gabe, and *p1* stands for race 2 *sensu* Cirulli and Alexander or race 1 *sensu* Gabe.

freedom for aggressiveness/virulence in the pathogen and therefore none for interaction.

Two other *Ht* genes in maize are known that are inherited independently of genes *Ht1* and *Ht2*. Gene *Ht3* was found in *Tripsacum floridanum* (Hooker, 1979), and gene *HtN* was found by Gevers (1975) in the Mexican maize variety Pepitilla. It would thus seem that there are three known degrees of freedom available in *Ht* resistance. but this does not help to distinguish between horizontal and vertical resistance, because the discovery of new genes for resistance in the host still leaves no degree of freedom for aggressiveness/virulence in the pathogen.

5.6 HOST AND PATHOGEN RANGES AS LIMITATIONS

It goes without saying that experimental findings hold only for the particular host and pathogen genotypes that are tested. In tests necessarily restricted by the conditions of the experiment, there can be no assurance that other genotypes lurking in the fields would not upset the findings. On the other hand when there has been experience of stability of resistance on a continental scale over long periods of time, as with many diseases of maize (see Section 8.8), one infers that both host and pathogen have had ample opportunity to vary, unless man introduces an abrupt variation, as he did when on a large scale he introduced *cms-T* cytoplasm for male sterility in maize. Limitations must be judged by the circumstances.

5.7 GLOSSARY

Durable Resistance is a term mainly used in England and best avoided. Durability and stability of resistance are agricultural ideals. The term stabilizing selection is standard in population genetics. Stabilizing selection helps to make resistance more stable. Even if one were willing to waive the convention that gives precedence to prior usage, one could still not say that durabilizing selection helps to make resistance more durable, because the verb, to durabilize, is unknown. In many contexts, the words stable and durable differ in meaning, but not in context with lasting resistance. See also ***Stable Resistance.***

General Resistance is an ill-defined term, possibly with the same meaning as race-nonspecific resistance.

Monogenic, Oligogenic, and ***Polygenic Resistance*** are of variable value. Monogenic and oligogenic resistance are acceptable terms in a genetic background. A polygenic character is not just a character conditioned by many genes. It is a character conditioned by many genes none of which has a large effect (or, alternatively, has an effect large in relation to nonheritable variation). There is yet no strict proof that any resistance to plant disease is polygenic, within the tight definition of the word (see also Chapter 8). Despite initial optimism, attempts to relate horizontal and vertical resistance to gene numbers have failed.

Partial Resistance is sometimes used as a synonym for horizontal resistance, but vertical resistance is often also only partial. This has been apparent ever since Stakman and his associates published lists of races of *Puccinia graminis* early this century. With reaction type 0 for full resistance and type 4 for full susceptibility, intermediate types can be taken to indicate partial resistance. With reaction type 4 indicating susceptibility, type 3 indicates some resistance, type 3− more of it, types 2+ or X still more, and so on. The list of Stakman *et al.* (1962) is rich in examples of partial resistance, and the resistance is undoubtedly vertical. There can be no objection to the use of the term partial resistance when it means incomplete resistance; the objection is to any built-in implication that incomplete resistance is necessarily horizontal.

Pseudospecificity occurs when a particular genotype of the pathogen increases as a result of an increased amount of inoculum or as the result of environmental conditions becoming more favorable for inoculation. (See also the ***Third Variable***.)

Race-Specific and ***Race-Nonspecific Resistance*** are terms widely used. Race-specific resistance is an acceptable term. Race-specific resistance is vertical resistance, but the converse is not necessarily true: Vertical resistance is not necessarily race-specific. If host variety 1 gave reaction types 2− and 2+ with pathogen races 1 and 2, respectively, and variety 2 gave types 2+ and 2− with the same races, the resistance would be vertical; but both varieties would be resistant to both races, so the resistance could hardly be called specific. Resistance which is vertical but nevertheless expressed against all known races is common in both fungus and bacterial disease; it is, e.g., shown by the barley variety Vada against *Puccinia hordei* (Clifford and Clothier, 1974), and by the cotton variety Albar 51 against *Xanthomonas malvacearum* (Cross, 1963). By subtraction, one concludes that race-nonspecific resistance is a mixture of horizontal resistance and that part of vertical resistance that is not race-specific.

Stable Resistance can be horizontal or vertical or both, but is also a concept outside the definitions of horizontal and vertical resistance, for two reasons. First, some of the most stable resistance cannot be classified as horizontal or vertical, because of inadequate degrees of freedom; examples were given in Section 5.5. Second, horizontal and vertical resistance are concepts within the frame of a two-variable system. Stable resistance should be seen within a multivariable system or at least a three-variable system. The next entry shows why.

Third Variable is the term proposed for the farmers' response. When a plant breeder releases a new cultivar with resistance to disease, farmers respond, quite rightly, by using that resistance. If the resistance is against a root disease, the response might be to shorten the crop rotation or to omit soil fumigation. If it is against a leaf disease, the response might be to economize on fungicides. The response might also be to start growing the crop where it was previously thought to be uneconomic, and so on.

Horizontal and vertical resistance are defined strictly within a two-variable system. The resistance itself almost inevitably introduces the third variable, which in turn introduces its own consequences. Consider fursarium wilt of tomatoes. Farmers in warm climates today grow tomatoes in a way they would not have dared do before the gene *I* became available; and the shortened rotations have led to the accumulation of race 2 (*sensu* Cirulli and Alexander) of the fungus, giving the race a *pseudospecificity* within the three-variable system.

It would, of course, be possible to subdivide the third variable almost endlessly, but we take the third variable essentially to mean more inoculum or inoculum which changed environmental conditions make more effective.

Tolerance implies that the host's loss of well-being as a result of infection is limited and less than it otherwise would have been. In the two-variable system which is the topic of this chapter, variations as a main effect are horizontal, whereas a host–parasite interaction means that the tolerance is vertical. When tolerance is incomplete it is integrated with other horizontal or vertical effects. When tolerance is practically complete, other effects become irrelevant. An example is the use of tolerant rootstocks to save citrus trees from the effect of infection by tristeza virus. Sometimes tolerated infection may even be beneficial, as when infection by potato virus X is tolerated by the host and reduces the susceptibility of potato tubers to a form of fusarium rot (Jones and Mullen, 1974). Epidemiologically, resistance and tolerance differ greatly. Adaptation pressure in the population of the pathogen tends toward the loss of vertical resistance in the host, this causes the bust in boom-and-bust cycles. But there is no corresponding reason why vertical tolerance should be lost. Because of this difference it is probably best to treat tolerance as a topic of symptomology rather than as one of resistance. However, the line between resistance and tolerance is badly smudged, and no shuffling of topics will remove the smudge.

Note Added in Proof

J. M. Perkins and A. L. Hooker (*Plant Dis.* **65,** 502–504, 1981) discuss a new race, race 3, of *Helminthosporium turcicum* virulent for maize genes *Ht*2 and *Ht*3 but not *Ht*1. The new race provides the degree of freedom needed to show that the *Ht* form of resistance is qualitative.

6

The Gene-for-Gene Hypothesis

6.1 INTRODUCTION

Flor's (1942) gene-for-gene hypothesis is perhaps this century's greatest contribution to plant pathology principle. This is recognized in the size of the literature of gene-for-gene disease. The general evidence has been reviewed by Flor (1971) himself, and by several others including Day (1974), Sidhu (1975), and Person and Ebba (1975). We need not here repeat much of a well-known literature. Instead, we emphasize elements which have not been widely discussed, especially the numerical implications and their chemical consequences.

In his work on flax rust, Flor was the first to study the genetics of both members of a host–pathogen system. From his experiments he concluded that for each gene determining resistance in flax (*Linum usitatissimum*) there was a specific and related gene determining pathogenicity in the rust fungus (*Melampsora lini*). In flax varieties possessing one gene for resistance to the avirulent pathogen, pathogenicity in a virulent race was conditioned by one gene in the fungus. In flax varieties possessing two, three, or four genes for resistance, pathogenicity was conditioned by two, three, or four genes in the fungus. The hypothesis, that for each resistance gene in the host there is a matching or reciprocal gene for pathogenicity in the fungus, is the simplest that fits these facts. The range of pathogenicity of a race of *M. lini* is determined by pathogenic factors specific for each resistance factor possessed by flax.

The hypothesis, that for every resistance gene in the host there is in disease a corresponding virulence gene in the pathogen, is an experimentally based principle. In the flax–flax rust system 26 resistance genes in the host have been matched with 26 virulence genes in the pathogen; for this system the principle is strongly based on experimental fact. In other systems the strength of the evidence varies from very high, particularly in the wheat–*Puccinia graminis* system, to rather poor. Table 6.1 lists various systems for which a gene-for-gene relationship has been demonstrated or suggested.

The systems listed in Table 6.1 are probably only a very small sample of reality. For example, only one species of powdery mildew, *Erysiphe graminis,* occurs in the list out of hundreds known; one must suspect that its unique feature

TABLE 6.1

Parasite–Host Systems for Which a Gene-for-Gene Relationship Has Been Shown or Suggested

Viruses	Tobacco mosaic virus-*Lycopersicon*
	Spotted wilt virus-*Lycopersicon*
	Potato virus X-*Solanum*
Bacteria	*Xanthomonas malvacearum*-*Gossypium*
	Rhizobium-Leguminoseae
Phycomycetes	*Phytophthora infestans*-*Solanum*
	Synchytrium endobioticum-*Solanum*
Ascomycetes	*Erysiphe graminis hordei*-*Hordeum*
	E. graminis tritici-*Triticum*
	Venturia inaequalis-*Malus*
Basidiomycetes	*Melampsora lini*-*Linum*
	Hemileia vastatrix-*Coffea*
	Puccinia graminis avenae-*Avena*
	P. graminis tritici-*Triticum*
	P. helianthi-*Helianthus*
	P. recondita-*Triticum*
	P. sorghi-*Zea*
	P. striiformis-*Triticum*
	Ustilago avenae-*Avena*
	U. hordei-*Hordeum*
	U. tritici-*Triticum*
	Tilletia caries-*Triticum*
	T. contraversa-*Triticum*
	T. foetida-*Triticum*
Deuteromycetes	*Cladosporium fulvum*-*Lycopersicon*
Nematodes	*Heterodera rostochiensis*-*Solanum*
Insects	*Mayetiola destructor*-*Triticum*
Angiosperms	*Orobanche*-*Helianthus*

is that a unique amount of genetic attention has been given to powdery mildew of wheat and barley. In general one expects to hear of gene-for-gene systems only where genetic research has been concentrated with the aim of developing resistant cultivars. Also, there must be resistance genes readily transferable. *Solanum tuberosum* has no *R* genes for resistance to *Phytophthora infestans*; a gene-for-gene system could be suggested only after *R* genes were introduced from *S. demissum* and only because *S. tuberosum* and *S. demissum* hybridize to give progeny which potato breeders thought worth studying. Eleven *R* genes were found while potato breeders were still optimistic about the use of *R*-gene resistance; now that the optimism has vanished, research on new *R* genes has practically stopped.

Table 6.1 gives other evidence of incompleteness. Whereas most classes of pathogen are represented in the table, the range of the host list is narrow and limited to a few crop plants. In particular they are crop plants in which vertical resistance has been used by plant breeders.

Table 6.2 illustrates the gene-for-gene system applied to *P. infestans* and potatoes with four dominant resistance genes, *R1, R2, R3,* and *R4,* singly and in combination. Specific virulence is given in terms of races: race 1 is virulent for gene *R1* but not *R2, R3,* or *R4*; race 1, 2 is virulent for genes *R1,* and *R2,* but not *R3* or *R4*; and so on.

6.2 BIOTROPHY AND GENE-FOR-GENE SYSTEMS

All the parasites listed in Table 6.1 are either essentially biotrophic (e.g., viruses, rusts, *Orobanche*) or biotrophic at least for some time after the start of infection (e.g., *Xanthomonas malvacearum, Phytophthora infestans, Venturia inaequalis*) (Vanderplank, 1978). Even the larvae of the Hessian fly (*Mayetiola destructor*) feed biotrophically, without destroying the nuclei of the host cells they are in contact with.

Here we have a hint that susceptibility in gene-for-gene systems is concerned not just with specific recognition but also with specific feeding. Resistance, on these lines, would be concerned not just with specific recognition but also with disrupting specific feeding; it is the ultimate and specific spanner thrown by the host plant into the works, the works in disease being the feeding of the pathogen.

Lest there be a misunderstanding let it be said that we do not equate all biotrophy with gene-for-gene systems. Gene-for-gene systems involve biotrophy, but the converse is not necessarily true. For example, *Agrobacterium tumefaciens,* the cause of crown gall, and *Plasmodiophora brassicae,* the cause of clubroot of crucifers, are biotrophic, but no evidence has yet been presented in the literature to suggest that they infect on a gene-for-gene system.

TABLE 6.2

International System of Designating Interrelationships of Genes and Races of *Phytophthora infestans*[a]

Genotypes	(0)	(1)	(2)	(3)	(4)	(1,2)	(1,3)	(1,4)	(2,3)	(2,4)	(3,4)	(1,2,3)	(1,2,4)	(1,3,4)	(2,3,4)	(1,2,3,4)
r	S	S	S	S	S	S	S	S	S	S	S	S	S	S	S	S
R1	R	S	R	R	R	S	S	S	R	R	R	S	S	S	R	S
R2	R	R	S	R	R	S	R	R	S	S	R	S	S	R	S	S
R3	R	R	R	S	R	R	S	R	S	R	S	S	R	S	S	S
R4	R	R	R	R	S	R	R	S	R	S	S	R	S	S	S	S
R1, R2	R	R	R	R	R	S	R	R	R	R	R	S	S	R	R	S
R1, R3	R	R	R	R	R	R	S	R	R	R	R	S	R	S	R	S
R1, R4	R	R	R	R	R	R	R	S	R	R	R	R	S	S	R	S
R2, R3	R	R	R	R	R	R	R	R	S	R	R	S	R	R	S	S
R2, R4	R	R	R	R	R	R	R	R	R	S	R	R	S	R	S	S
R3, R4	R	R	R	R	R	R	R	R	R	R	S	R	R	S	S	S
R1, R2, R3	R	R	R	R	R	R	R	R	R	R	R	S	R	R	R	S
R1, R2, R4	R	R	R	R	R	R	R	R	R	R	R	R	S	R	R	S
R1, R3, R4	R	R	R	R	R	R	R	R	R	R	R	R	R	S	R	S
R2, R3, R4	R	R	R	R	R	R	R	R	R	R	R	R	R	R	S	S
R1, R2, R3,	R	R	R	R	R	R	R	R	R	R	R	R	R	R	R	S

[a] Here, S stands for susceptible and R for resistant.

6.3 POSSIBLE GENE DUPLICATION

It is possible that the future may show the need for changes in detail without necessarily changing the essence of the hypothesis.

There is experimental evidence for two operative virulence genes where only one is to be expected. Ebba and Person (1975) found evidence that in the barley-smut system the virulence of *Ustilago hordei* on barley cultivars Keystone and Himalaya can be determined by two recessive genes at either of two genetic loci. One must accept the tentative possibility that duplicate genes are here operating on a gene-for-gene basis. Statler and Zimmer (1976), working with the flax–flax rust system, investigated the segregation for pathogenicity among S1 cultures derived from selfing race 370 of *Melampsora lini*. Race 370 is avirulent for the gene *L1* of flax. Its selfed progeny gave 22 cultures avirulent and 3 virulent for the gene *L1*, from which ratio Statler and Zimmer suggest a two-gene explanation.

Caution prompts one to await further elucidation and extension of these findings, but meanwhile it can be said they are not unexpected. The systems concerned are eukaryotic, and the genes cannot come into contact directly, i.e., there is no contact as DNA level. Gene products are involved. If there was gene duplication, with the duplicates having the same recognition system, it would probably pass unnoticed experimentally while the duplicates were tightly linked but would become noticeable if translocation broke the linkage.

6.4 MULTIPLE ALLELES WITH THE SAME RECOGNITION SYSTEM

There is evidence that there can be more than two alleles at a locus, all of them with the same recognition system.

Watson and Luig (1968) selected a culture of *Puccinia graminis tritici* which produced gray-brown uredospores. The purpose of using specially colored spores was to insure against undetected contamination. The culture was avirulent for wheat plants with the gene *Sr11*: the reaction type with these plants was 0, indicating high resistance. Starting with a single spore they increased the culture on plants of a susceptible variety, and then retested it on plants with the gene *Sr11*. One pustule was found giving an intermediate reaction type, X=. This in turn was increased on plants of a susceptible variety, and then again retested on plants with the gene *Sr11*. They then found a pustule with a mildly susceptible reaction 3. (The culture from it gave a 2+, 3c, and 3 type reaction of the same leaf.) Wheat plants with the gene *Sr11* give a 4 reaction with fully virulent cultures, so here we have evidence for four reaction types, 0, 2=, 3, and 4 on plants with the gene *Sr11*. Four cultures with different reaction types would be recognized by their behavior on plants with gene *Sr11*.

Virulence for the gene *Sr9g* is very unstable, and progressive increases or decreases occur naturally (Luig, 1979). Indeed, progressive levels of virulence and avirulence now appear to be typical of most *Sr* loci in wheat (McIntosh, 1977). There are five known levels of interaction involving virulence and avirulence for the gene *Sr15* (McIntosh, 1977). In oats Luig and Baker (1973) noted that there were two reaction types (1 and 2+) representing resistance when the variety Rodney, with the gene *Pg4,* was inoculated with *P. graminis avenae,* and they ascribed this to multiple alleles for virulence in the pathogen.

Evidence for multiple alleles with the same recognition system also comes from experiments in which a culture was trained to become more pathogenic. Whereas in the experiments just considered the mutations occurred while the pathogen was multiplied on plants of a susceptible variety, training is effected on plants of a resistant variety. Reddick and Mills (1938) and de Bruijn (1951) trained *Phytophthora infestans* to become gradually more virulent. They inoculated plants of a resistant variety from which they managed to obtain a few spores. These were used to start a culture which was slightly more virulent. Successively plants of a resistant variety were inoculated, successively more and more spores were obtained, until finally the culture was fully virulent.

The gene-for-gene hypothesis must probably be elaborated to read that for every resistance gene in the host there is a corresponding and specific virulence gene in the pathogen with an indefinite number of alleles of varying virulence. Ecologically, in farming practice, the elaboration may be of little importance, because on plants with a resistance gene the pathogen would tend to mutate to maximum virulence compatible with fitness. But there are two theoretical implications.

First, the system of classifying isolates simply as virulent or avirulent is, despite complication by occasional intermediate reactions, preferable to the old system of ''standard'' races with seven or more reaction categories ranging from 0 to 4, with various plus and minus amendments. The old system was meaningless unless there was stability. Imagine, e.g., that a wheat variety with gene *Sr11* had been one of the standard differential varieties. Watson and Luig's experiments would have added two new standard races even although the number of spores available for mutation in their experiment was infinitesimally small compared with the number produced in a stem rust epidemic.

Second, in any biochemical interpretation of the gene-for-gene hypothesis one must look for molecules large and flexible enough to accommodate a moiety concerned with recognition and a moiety subject to variation without changing recognition. This matter is taken up again, against a different background, in Section 6.13.7.

6.5 PSEUDOALLELES WITH DIFFERENT RECOGNITION SYSTEMS

In Flor's discoveries of the flax–flax rust gene-for-gene system there were 26 resistance genes in five loci, the *K, L, M, N,* and *P* loci. The genes in a locus were provisionally taken to be allelic because they failed to show recombination among the F_2 and F_3 progeny. But they could equally have been tightly linked pseudoalleles or duplicates. Flor (1965) himself discovered them to be so; he detected rare recombinations between genes from each of the *L, M,* and *N* groups in flax. So too Saxena and Hooker (1968), working with the maize–maize rust system, detected rare recombinations between genes from the *Rp1* group in maize; and Dyck and Samborski (1970), working with the wheat–leaf rust system, combined *Lr14a* and *Lr14b* in a single line of wheat. There are still some details of chromosome structure to be worked out (Shepherd and Mayo, 1972) and many more recombination experiments are needed. But on the evidence we accept with a high degree of probability that we are dealing not with alleles but with closely linked and separately recognized duplicates.

The distinction is important. In Section 2.19 we referred to the possibility of creating a supergene by combining two genes from the wheat "locus" *Sr9* in one line. Pseudoallelism would allow that. Moreover, in relation to biochemical explanations of the gene-for-gene system it is relevant to know that resistance genes and susceptibility genes correspond numerically. That is, in flax, for example, it is relevant to know that with 26 resistance genes there are 26, not 5, susceptibility genes. In this way we get numerical correspondence, via the gene-for-gene hypothesis, between susceptibility genes in the host and virulence/avirulence genes in the pathogen.

On the pathogen's side, less evidence has been collected about pseudoallelism. The locus in *Melampsora lini* corresponding to the *P* locus in flax has been studied in some detail; and the evidence is that virulence for flax rust resistance genes *P, P1, P2,* and *P3* is conditioned by separate but closely linked genes (Flor, 1960; Day, 1974, quoting unpublished work by Lawrence). Because spores are produced abundantly, recombination is unlikely to be a problem for the pathogen; and the extent of virulence pseudoallelism in pathogens may perhaps be greater than the evidence now suggests.

6.6 THE QUADRATIC CHECK VERSUS THE MINIMUM TEST FOR THE HYPOTHESIS

Recurrent in the recent literature are references to the quadratic check which is supposed to be a test for whether a gene-for-gene relation exists. The essential point about the quadratic check is that it concerns a single gene for resistance in the host and a single gene for virulence in the pathogen. Against this, Flor's hypothesis states that for each gene for resistance in the host there is a corresponding gene for virulence in the pathogen. Obviously in order to test this hypothesis there must be at least two genes for resistance in the host and two for virulence in the pathogen, whereas the quadratic check provides only one. That is, in the quadratic check there are zero degrees of freedom between resistance

TABLE 6.3

An Illustration of the Quadratic Check[a,b]

	Host	
Pathogen	*R1R1*[c]	*r1r1*
V1V1	Resistant[d]	Susceptible
v1v1[b]	Susceptible	Susceptible

[a] Host reaction when alleles *R1* for resistance and *r1* for susceptibility at a single locus in the host interact with alleles *V1* for avirulence and *v1* for virulence at a single locus in the pathogen

[b] With only one resistance gene *R1* in the host and one virulence gene *v1* in the pathogen, there are no degrees of freedom to test for a gene-for-gene relation.

[c] *R* is assumed to be dominant and *R1R1* can be replaced by *R1r1; v* is assumed to be recessive and *V1V1* can be replaced by *V1v1*.

[d] The host is resistant.

genes and zero between virulence genes; and no possibility for checking exists. The minimal test requires two resistance genes and two virulence genes, which gives one degree of freedom for resistance genes and one for virulence genes, and therefore one for the interaction between resistance genes and virulence genes.

Tables 6.3 and 6.4 compare and contrast the quadratic check with the minimum needed to test the gene-for-gene hypothesis.

Table 6.4 keeps to the point of the gene-for-gene hypothesis. The quadratic

TABLE 6.4

The Minimum Test for a Gene-for-Gene Relation[a,b]

	Host	
Pathogen	*R1R1*[c]	*R2R2*
v1v1	Susceptible[d]	Resistant
v2v2	Resistant	Susceptible

[a] Host reaction when resistance genes *R1* and *R2* at two loci in the host interact with virulence genes *v1* and *v2* at two loci in the pathogen

[b] With two resistance genes in the host and two virulence genes in the pathogen there is one degree of freedom to test resistance-virulence interaction.

[c] Resistance is assumed to be dominant, and *R1R2* can be replaced by *R1r1* and *R2R2* by *R2r2*. Virulence is assumed to be recessive.

[d] The host is susceptible.

check illustrated in Table 6.3 wanders off the point. The hypothesis is about resistance genes in the host and virulence genes in the pathogen. Susceptibility genes in the host and avirulence genes in the pathogen do not enter the enunciation of the hypothesis. (''For each gene conditioning resistance in the host there is a related and specific gene conditioning virulence in the pathogen.'') Susceptibility genes and avirulence genes enter the quadratic check as irrelevancies.

6.7 SUSCEPTIBILITY IS SPECIFIC

When a host plant with a resistance gene is inoculated with a pathogen having the related and specific virulence gene, the plant is susceptible. To refer to Table 6.4, when a plant with the resistance gene *R1* meets a pathogen with its related and specific virulence gene *v1*, the host is susceptible; when a plant with the resistance gene *R2* meets a pathogen with the related and specific virulence gene *v2*, the plant is susceptible. This is what the gene-for-gene hypothesis clearly states, with no ambiguity whatsoever. Specificity, as defined in Flor's gene-for-gene hypothesis, resides in susceptibility.

There have been attempts, started by Ellingboe (1976), to argue that specificity in host–parasite systems resides in resistance alone. This is so flatly contrary to the unequivocal statement of Flor's hypothesis, that no scope for debate remains. Something more about specificity, in a somewhat different context, is taken up in Section 6.14.

Any biochemical explanation of the gene-for-gene hypothesis must include an explanation of how susceptibility comes to be specific. As it happens, the explanation we shall suggest covers both specific susceptibility and specific resistance; but this does not detract from the fact that the primary target is to explain specific susceptibility.

The phytoalexin theory that ignores specific susceptibility is inapplicable to gene-for-gene disease, even though the theory originated with potato blight, which is thought to be a gene-for-gene disease.

6.8 THE NUMERICAL AND CHEMICAL IMPLICATIONS OF THE HYPOTHESIS

Consider the three rust diseases of wheat. Wheat has more than 30 known *Sr* genes for resistance to *Puccinia graminis,* more than 20 known *Lr* genes for resistance to *P. recondita,* and more than 10 known genes for resistance to *P. striiformis*. To these one might add the number of genes for resistance to powdery mildew and other wheat diseases listed in Table 6.1. On a conservative estimate one must allow for at least 60 genes for resistance and 60 for susceptibility to various diseases. That is, there are potentially at least 2^{60} ($\sim 10^{18}$) phenotypes of wheat distinguishable by one or other pathogen or race of pathogen.

The previous paragraph gives two numbers: 60 and 10^{18}. Both are relevant. Each gives a clue to the chemistry of gene-for-gene relations.

The number 10^{18} is so great that it is incomprehensible. One can only make a comparison. If one goes back to the creation of the universe starting, the astronomers say, with the Big Bang, 10^{18} exceeds the number of minutes that on the best estimates have ticked away ever since the beginning of time. Such a number tells us that there cannot possibly be an explanation of gene-for-gene relations based on *quantitative* differences in substances, i.e., on different *amounts* of substances, be they phytoalexins, auxins, gibberellins, cytokinins, ethylene, lignin, cutin, or any other plant or microbial constituent one cares to name.

The number 60 or, better, $60 \times 2 = 120$, prescribes the number of *qualitative* differences one must be prepared to account for, the qualitative differences being differences in *sort* that host and pathogen can recognize. Qualitative variation of this magnitude, coupled with recognition, is possible only with poly compounds: polynucleotides (DNA and RNA), polypeptides (proteins), and polysaccharides. In these compounds variation in the order, arrangement, and number of the subunits permits variants to exist in endless permutations and combinations.

All heritable variation is stored in the DNA (the RNA viruses apart), and for the purpose of the present discussion variation can be considered to pass along the sequence DNA, RNA, protein (enzyme), and polysaccarides. With this sequence possible in both host and pathogen, where in the sequence does host–pathogen recognition lie? At what point, or points, in the sequence do the molecules carrying the variation of host and pathogen come together and recognize one another? Coming together implies close contact within the range of chemical bonds. That is the central biochemical problem of gene-for-gene relations.

6.9 THE AXENIC CULTURE FALLACY

Consideration of the numerical implications of the gene-for-gene hypothesis leads straight to the study of macromolecules: polynucleotides, proteins, and polysaccharides. But the involvement of macromolecules in biotrophy has been disputed because of a fallacious argument about axenic culture. This must be corrected.

Williams *et al.* (1966) succeeded in culturing *Puccinia graminis tritici* axenically on a simple medium. Since than *P. graminis tritici* has been cultured again by Green (1976c) and Bose and Shaw (1974), *P. graminis avenae* and *P. graminis secalis* by Green (1976c), *P. recondita* by Singleton and Young (1968) and Green (1976c), *P. coronata* by Green (1976c), *Uromyces dianthi* by Jones

(1973), *Melampsora lini* by Turel (1969) and Bose and Shaw (1974), *Cronartium fusiforme* by Hollis *et al.* (1972), and *C. ribicola* by Harvey and Grasham (1974).

Axenic cultures are abnormal. To use *P. graminis tritici* as an example, the dikaryotic uredospores used as inoculum give rise to two sorts of axenic colonies. There are colonies with binucleate cells, which after a long interval start growing comparatively fast but soon stop growing; and there are colonies with mononucleate cells which grow very slowly but persist better. Some isolates cannot easily synthesize new nuclear material, which is compensated for by the continual migration of old nuclei into younger mycelium.

The fallacious argument in the literature (see e.g., Chakravorty and Shaw, 1977) is that because these pathogens can grow *in vitro* and some can sporulate in the absence of the host and in completely defined, simple media, it is unnecessary to assume macromolecular transfer from the host to the pathogen. The argument implies that any particular organism has one and only one method of feeding in all circumstances, and misses the fact that in nature, let alone *in vitro,* organisms adapt a changing nutrition to changing circumstances. Consider *Venturia inaequalis*. Young apple leaves are infected, and at the start the growth of the fungus is biotrophic. Later in the season the mycelium which was initially subcuticular invades the mesophyll cells of the leaf and growth becomes necrotrophic. Still later, the leaves fall, perithecia develop in the mesophyll, and their growth is clearly saprotrophic. Biotrophy, necrotrophy, and saprotrophy are used in succession. Or consider *Phytophthora infestans*. For the first few days after infection, growth is biotrophic, but about the time sporulation begins cells are killed, and in old lesions a zone of biotrophic infection surrounds a sporulating zone with much necrosis. If we accept, as we must, that *V. inaequalis* and *P. infestans* feed differently in their biotrophic and necrotrophic phases, for necrotrophy is the antithesis of biotrophy, then we must accept that the biotrophic nutrition of *Puccinia graminis* in a wheat leaf is different from its nutrition *in vitro*. Moreover, can we believe that there would be stem rust epidemics if *P. graminis* on wheat was the same slow-growing, genetically deformed runt that it is in even the best artificial media that have yet been devised?

The discovery by Williams *et al.* (1966) that *P. graminis* can be made to grow on a simple medium was dramatic and one of the highlights of mycology. But it throws little light on how *P. graminis* feeds in wheat. Any expectations about that have long since been dispelled.

6.10 DNA

Does gene-for-gene disease involve DNA-for-DNA contact? For eukaryotic parasites this seems improbable, because there is no evidence for exchange of genetic material during infection. For prokaryotic parasites the scope is wider. *Agrobacterium tumefaciens* incorporates plasmid DNA into the host's genetic system to cause crown gall. But at present there is no evidence that *A. tumefaciens* operates on a gene-for-gene system. As to viruses, the enzyme reverse transcriptase discovered in animal tumors in 1970 makes it possible for RNA to be transcribed into DNA for incorporation into the host's genetic system. But the RNA viruses listed in Table 6.1 are not tumor viruses and there is no evidence that they are multiplied together with the chromosomes during cell division.

In summary, there is as yet no evidence that gene-for-gene relations involve DNA-for-DNA contact, and we can almost certainly rule the possibility of such contact out when the pathogen is eukaryotic.

6.11 RNA

There are both qualitative and a quantitative aspects to the involvement of RNA in gene-for-gene disease.

Qualitatively, there is the matter of specificity. It is highly probable that variation in the DNA, axiomatic in the gene-for-gene hypothesis, reaches the RNA. But specific variation in RNA is also a corollary to specific variation in protein and specific variation in polysaccharide. The discovery of specific RNA in host and pathogen does not necessarily imply that host–pathogen recognition is at RNA level; it could equally well be at protein or saccharide level.

Rohringer *et al.* (1974) found evidence for a gene-specific RNA determining resistance to stem rust in wheat; but the technique did not allow the separate determination of mRNA, so we are still in the dark about where in the sequence of synthesis RNA fits.

Quantitatively, there is evidence for an increase in the RNA content of both host and pathogen. The literature, starting with Allen (1926), is large and has been reviewed by Samborski *et al.* (1978), among others. In wheat stem rust, the most widely studied example, the host nucleus in compatible host–pathogen pairs swells after the invasion of the protoplast by the fungus, and this swelling is accompanied by a swelling of the nucleolus and an increase in activity of the RNA. The increase of activity is greater in incompatible host–pathogen pairs, as was shown by Von Broembsen and Hadwiger (1972) for flax rust. One infers that metabolic disruption is greater when the processes lead to wounding and repair, as shown by hypersensitive responses, than when they lead to biotrophy. This is not unexpected; the very term hypersensitivity foretells the evidence.

There is no evidence for host RNA–pathogen RNA associations that would explain gene-for-gene relations, which is not to say that suitable models will not one day be presented. Other models, as for tRNA–protein, may yet come.

Meanwhile, on the evidence available, it seems that RNA is more likely to be involved as messenger than as the contact point between host and pathogen.

6.12 PROTEIN

There is much evidence for a synthesis of protein in host tissues as a result of infection. For example, Pozsár, Kristev, and Király, quoted by Goodman *et al.* (1967), incorporated ^{35}S-cysteine into the protein fraction of bean leaves infected with *Uromyces phaseoli* and found about twice as much protein at infection centers as elsewhere in the leaf. As Farkas (1978) has pointed out, the limits set by the present-day rapid separation techniques are two orders of magnitude lower than the actual number of proteins to be separated. Techniques which favor the detection of nonspecific changes in protein are therefore unlikely to help us understand gene-for-gene systems.

More specific information has been given by Macko *et al.* (1968), Seevers and Daly (1970), Daly *et al.* (1974), Seevers *et al.* (1971), and Daly (1972) about protein in the form of peroxidase. They studied wheat stem rust reactions controlled either at the *Sr6* or *Sr11* locus. Infection caused peroxidase concentrations to increase. The increase was relatively small in compatible (susceptibility) reactions, and tended to be followed by a decrease when once sporulation began. This is consistent with a mild reaction followed by strong biotrophic feeding to produce spores. The increase was much stronger in incompatible (resistance) reactions. With proteins, as with RNA, the metabolic disturbance associated with the induction and repair of necrotic lesions is great. Of special interest were results from separating various isozymes of peroxidase. Of the 14 isozymes detected in both healthy and infected leaves, increases of only one, isozyme 9, were associated consistently with the development of a resistance reaction. The reactions of the gene *Sr6* are temperature sensitive, with a resistance reaction at 20° but a susceptibility reaction at 25°C. When, some days after infection, plants with a high induced peroxidase activity associated with resistance at 20° were transferred to 25°C they reverted to complete susceptibility. However, the disease-induced activity of isozyme 9 did not fall. These data, they point out, suggest that the association of peroxidase isozyme 9 with resistance is a consequence of, not a determinant in, resistance. At least it seems clear that the metabolic processes associated with biotrophy and those associated with incompatibility should be studied apart.

Data similar to that for peroxidase have been given by Henderson and Friend (1979) for phenylalanine ammonia lyase (PAL). After infection by *Phytophthora infestans* the PAL content of potato tuber discs increased, sharply when the potato variety was resistant, less sharply when it was susceptible. But on the third day after infection the PAL content in the tuber discs of susceptible varieties decreased, as though PAL was being consumed.

6.13 THE PROTEIN-FOR-PROTEIN HYPOTHESIS

Proteins can store variation freely; indeed, most of the variation in the DNA of structural genes survives to be reflected in the amino acid residues of proteins. Proteins recognize one another precisely, as in many antigen–antibody systems or in polymerization. These two attributes, variation storage and mutual recognition, are essential for a biochemical explanation of the gene-for-gene hypothesis; and a protein-for-protein hypothesis was proposed by Vanderplank (1978).

The hypothesis states that in gene-for-gene disease the mutual recognition of host and pathogen is not by the genes themselves but by their coded proteins. In disease, i.e., in susceptibility, the proteins of the pathogen polymerize each with its partner host protein, specifically and with complete mutual recognition. Failure of any to polymerize means resistance to disease. The protein surfaces buried during polymerization resemble the antigenic determinants on the surface of globular proteins: patches of amino acid residues that determine antigenic binding and specificity in serology. Polymerization involves similar binding and specificity, with equal precision. It is not implied that all antigenic determinants are polymerization determinants, but it is believed that all polymerizing surfaces are, at least in part, antigenic. The site of polymerization in the host cell is at or near the host–pathogen interface.

There is substantial evidence for the hypothesis which will be presented. The essential difficulty of proof is the difficulty of reproducing *in vitro* what is believed to happen *in vivo*. It is an integral part of the hypothesis that the host–pathogen polymers should be loose, i.e., that the dissociation (equilibrium) constant should be relatively large. Loose polymers would explain the temperature effects discussed in the next few sections; they are also indicated for free-energy reasons (see Chapter 7). Loose polymers concentrated near the interface would inevitably dissociate, for mass-action reasons, when the solvent is diluted by sap from the rest of the cell, as would happen when sap is expressed in the laboratory. Finer methods of analysis may detect the relevant monomers in expressed sap, but they are unlikely to detect the polymers directly, because it is unlikely that the polymers can exist there. Sap concentration or other processes of extraction might help, but only against the background that protein polymerization is very sensitive to pH and to other substances in the solvent (see Section 7.4) For the time being one must therefore study indirect evidence. It comes from various sources. Some of it is specific, or appears to be specific, for proteins; some of it is apt for proteins but could be relevant elsewhere as well. It is set out in the next few sections in the form of tests, so that its merit can clearly be judged against experimental fact.

6.13.1 First Temperature Test: The Effect of Temperature on Resistance

If temperature changes resistance, it is in the direction of resistance at lower temperatures changing to susceptibility at higher temperatures. This follows from polymerization being an endothermic process (Oosawa and Asakura, 1975; Lauffer, 1975), and therefore promoted by increased temperature (see Chapter 7). The evidence for the change in gene-for-gene disease from resistance to susceptibility with increasing temperature is massive: the test is clearly met and passed.

A commonly quoted example is illustrated in Table 6.5, based on data of Watson and Luig (1968). The gene *Sr6* in wheat is effective against *Puccinia graminis tritici* at a temperature of 15°–20°C; but when the temperature is raised to 25°C or more, the resistance fades into susceptibility. Before Table 6.5 is considered in detail, reference will be made to some of the literature.

Johnson (1931) found that several varieties of durum wheat resistant to some races of *P. graminis tritici* at 16°C became susceptible to these same races at 24°C. Mahomet (1954) found that *P. graminis tritici* race 139 gave a type 2 reaction on Marquis and Kota wheats at 16°–22°C, but types 3–4 at 29°–32°C. Thatcher wheat grown with ammonium sulfate as a source of nitrogen was resistant to *P. graminis tritici* race 56 at 18°–24°C, but susceptible at 27°C (Daly, 1949). Over the same range of temperature the change was from type X (mesothetic) to susceptible when nitrate replaced the ammonium salt. Of the

TABLE 6.5

Reaction of Eureka Wheat, with Gene *Sr6* When Inoculated with Six Avirulent Cultures and a Virulent Culture of *Puccinia graminis tritici*[a]

	Temperature (°C)			
Culture	15	18-21	21-24	24-27
126-6,7	R	R	X	S
21-2	R	X	S	S
21-2.3,7	R	X	S	S
17-1,2,3,7	X	S	S	S
21-1,2,3,7	X	S	S	S
34-1,2,3,7	X	S	S	S
Virulent	S	S	S	S

[a] From data of Watson and Luig (1968). R, resistant; S, susceptible; X, intermediate.

wheat varieties he tested Bromfield (1961) found that 20 inoculated with race 17, 12 with race 38, and 15 with race 56 of *P. graminis tritici* were resistant at 21°C but susceptible at 25°–29°C. Watson and Luig (1966) found that the gene *Sr15* was effective only up to temperatures of 18°–21°, and that the variety Celebration lost its resistance to race 21 of *P. graminis tritici* at 24°–27°C. Gough and Merkle (1971) found that two genes in the variety Agent for resistance to stem rust race 111-SS2 and two in Agrus, including one from *Agropyron elongatum* were effective at 20°C, moderately effective at 25°C, but ineffective at 30°C. The gene *Sr5* in a Marquis background was effective against stem rust race 21-4, 5 at 24°C but not 30°C, and genes *Sr8* and *Sr9b,* when heterozygous, became ineffective against this race at high temperatures (Luig and Rajaram, 1972). The gene *SrFr2* is also sensitive to temperature, becoming ineffective at 27°C (Sanghi and Luig, 1974).

To turn now to *P. graminis avenae* Waterhouse (1929), Gordon (1933), and Newton and Johnson (1944) found the resistance of Joanette oats to be sensitive to temperature. The variety was resistant to stem rust races 1, 3, 4, and 5 at 12°C, but susceptible to races 3 and 4 at 24°C, and to all four races at 28°C. Newton and Johnson also found that some Hagira oat crosses were resistant at low temperatures but susceptible at 27°C, and that the variety Sevenothree, normally resistant, gave an intermediate reaction in the heat of summer. Hingorani (1947), quoted by Hart (1949), found Jostrain oats to be intermediate in resistance to stem rust race 2 at 16°C but susceptible at 21°–24°C; it was also moderately resistant to races 5 and 10 at 18°C, less resistant at 21°–24°C, and susceptible at 29°C. Ibrahim (1949), also quoted by Hart (1949), found that the oat variety Garry and some Hagira derivatives were normally resistant to stem rust race 6 but became susceptible at 29°C. Roberts and Moore (1956) found four oat varieties to be resistant to stem rust at 24°C but not at 29°C. Martens *et al.* (1977, 1979), when it was possible to study the reaction of stem rust resistance genes individually, found that genes *Pg3* and *4* were effective only up to 20°C; *Pg8, 9, 13,* and *15* were variably effective at 25°C but ineffective at 30°C; and only genes *Pg1* and *2* were effective up to the highest temperature studied, 30°C.

Studying wheat leaf rust caused by *Puccinia recondita,* Johnson and Schafer (1965) found that the wheat varieties Colotana, Lageadinho, Frontana, and La Prevision 25 were most resistant to leaf rust races 1 and 2 at 16°C, less resistant at 21°C, and susceptible at 27°C. Rajaram *et al.* (1971) found the variety W3301 to be resistant at 15°–18°C, but intermediate at 24°–27°C. Jones and Deverall (1977) found that the gene *Lr20* was effective at 20.5°C, partially effective at 26°C, and ineffective at 30.5°C. It has been suggested (McIntosh, 1977) that the gene *Lr20* is the same as *Sr15* which is also sensitive to temperature.

Resistance in wheat to stripe (yellow) rust caused by *Puccinia striiformis* is also sensitive to temperature. Pochard *et al.* (1962) found that some French wheat varieties in the field became susceptible to stripe rust as the temperature increased. Strobel and Sharp (1965) found the wheat varieties Rego and Nord Desprez to be resistant at night/day temperatures of 2°/18°C but susceptible at 15°/24°C (see Table 6.6). Fuchs and Hille (1968) also found the variety Nord Desprez to be resistant at low temperatures but susceptible at high temperatures to the same races. Nord Desprez has the resistance gene *Yr3*. Beaver and Powelson (1969) found that many wheat varieties (Chinese 166, Dippes Triumph, Etoile de Choise, Gaines, Leda, Omar) were resistant at night/day temperatures of 2°/18°C, but susceptible at a constant temperature of 18°C.

There is an early literature of temperature sensitivity in crown rust of oats caused by *Puccina coronata*. Peturson (1930) found Red Rustproof oats to be moderately resistant to race 4 at 14°C, but susceptible at 21°–25°C; the varieties Green Mountain, White Tartar, and Green Russian were resistant to race 7 at 21°C, but susceptible at 25°C. Rosen (1955) found oat lines that were resistant to races 45, 47, and 101 at 16°C but susceptible at 21°–24°C. Simons (1954) found the varieties Appler and Mo. 0-205 to be more susceptible at 25° than at 15°C, especially after the young seedling stage. Zimmer and Schafer (1961) found the variety Glabrota to be resistant to race 263 at 16°C, intermediate at 21°C, and susceptible at 27°C.

To turn from the rusts to other fungi, a recessive gene in the wheat varieties Hope and Renown is effective against relevant races of *Erysiphe graminis tritici* at temperatures up to 20°C but ineffective at 24°C (Futrell and Dickson, 1954; McIntosh *et al.* 1967). The potato blight resistance gene *R3* is ineffective in tubers at 20°C (Zacharius *et al.*, 1976).

There is evidence of temperature sensitivity in resistance to two of the bacteria listed in Table 6.1. Cotton plants (*Gossypium hirsutum*) with the genes *b* and *BN*

TABLE 6.6

Wheat Stripe (Yellow) Rust: The Number of Wheat Plants Classified as Resistant (R) or Susceptible (S) to *Puccinia striiformis* at Two Different Night/Day Temperature Profiles, 2°/18° and 15°/24°C[a]

	Reaction at 2°/18°C		Reaction at 15°/24°C	
Wheat	R	S	R	S
Rego	146	0	7	100
Rego × Lemhi F_1	0	8	0	13
Rego × Lemhi F_2	140	310	2	151

[a] From data of Röbbelen and Sharp (1978). In order to evaluate their evidence, plants which they recorded as having infection types 00, 0−, 0, 1−, 1, and 2 are here shown as resistant (R), and those they recorded as having infection types 3 and 4 are here shown as susceptible (S). The wheat variety Lemhi was susceptible, with reaction type 4, at both temperature profiles.

are resistant to *Xanthomonas malvacearum* at temperatures of 25.5°C by day and 19°C by night, but susceptible when the day temperature is raised to 36.5°C (Brinkerhoff and Presley, 1967). An effect of still higher temperature is discussed in a later section. *Rhizobium* needs special comment.

Rhizobium-legume relations differ from disease relations in that symbiosis beneficial to the "host" makes resistance alleles harmful and selected against. Nevertheless, resistance does occur; and one example simulates plant disease systems and is reminiscent of the gene *Sr6* system in wheat illustrated in Table 6.5. Lie *et al.* (1976) found that the pea cultivar Iran was resistant to the "avirulent" *Rhizobium* strains PRE, PF2, and 313 at 20°C, and this resistance was conditioned by a single dominant gene. At 26°C the cultivar was susceptible to all these strains. To the "virulent" strain 310 it was susceptible at both 20°C and 26°C (see Table 6.7). Plants inoculated with the avirulent strain 313 and kept for 3 days at 26°C before being transferred to 20°C were susceptible, just as in Antonelli and Daly's (1966) experiments wheat plants with the gene *Sr6* inoculated with an isolate of *Puccinia graminis tritici* avirulent for gene *Sr6* at 20°C gave a susceptible reaction if kept for 4 days at 25°C before being transferred to 20°C. There is yet another resemblance. The optimum temperature for the infection of susceptible varieties of wheat with stem rust is, as judged by records of stem rust epidemics, about 20°C, and at higher temperatures epidemics are on the wane. So too in most susceptible pea cultivars 20°C is about optimal for nodulation and 26°C well above the optimum. Thus, in both systems resistance is effective at temperatures optimal for infection, and ineffective at supraoptimal temperatures.

In terms of feeding, the difference between mutualistic symbiosis and antagonistic symbiosis might not be so great as would superficially appear.

TABLE 6.7

The Effect of Temperature on Root Nodule Formation by Four Strains of *Rhizobium* in the *Pisum sativum* Cultivar Iran[a]

	Temperature (°C)	
Strain	20	26
PRE	R[b]	X
PF2	R	S
313	R	S
310	S	S

[a] From data of Lie *et al.* (1976).

[b] R, few or no nodules; X, intermediate number of nodules; S, abundant nodulation.

Rhizobium fixes nitrogen as ammonia which the host plant promptly steals from the bacteroids, leaving the bacterium as dependent on its host as parasitic bacteria are.

Because resistance to nodulation would harm legumes by preventing infection with *Rhizobium,* one assumes that resistance genes of the sort found by Lie *et al.* (1976) would occur only in climates with hot summer soils, as in Iran. What would be a deleterious mutation in cool soil would be a neutral mutation in a hot soil.

Of the virus diseases listed in Table 6.1 only tomato mosaic seems to have been studied in relation to the effect of temperature on the effectiveness of resistance genes. The combination, tobacco mosaic virus on tomato, has been described as being in a gene-for-gene system (Pelham, 1966). Cirulli and Ciccarese (1975) collected isolates of tobacco mosaic virus from field-grown plants of tomato, tobacco, pepper (*Capsicum annuum*), and *Solanum nigrum,* and tested them on tomato plants carrying the resistance genes *Tm1*, *Tm2,* and *Tm2a,* singly and in combination. Resistance was reduced at higher temperatures. Averaged for the three resistance genes, heterozygous resistance was effective against 56% of the isolates at 17°, 38% at 22°; 19% at 26°, and 14% at 30°C. The corresponding figures for homozygous resistance were 88%, 79%, 72%, and 61%.

This section opened with the statement that *if* temperature changes resistance, the change is in the direction of resistance at lower temperatures and susceptibility at higher temperatures. Observe the conditional *if.* It is not implied that all resistance is unstable in relation to temperature. It is implied that there is no essential thermodynamic difference in effect between temperature-stable and temperature-unstable genes. Polymerization of protein is enhanced by increased temperature along the scale of absolute temperatures; but we can study only that part of the scale that lies between the mimimum and maximum cardinal temperatures for disease. Indeed, temperature is usually studied over only that part of the scale which is normally maintained in the greenhouses and laboratories of experiment stations. The difference between the temperature-unstable wheat stem rust resistance gene *Sr6* and the temperature-stable gene *Sr11* is (according to the hypothesis) that in the *Sr6* system protein polymerization occurs before the maximum cardinal temperature for wheat stem rust (approximately 30°C) is reached, while in the *Sr11* system the maximum cardinal temperature is reached before polymerization occurs. Studies of the *Sr6* and *Sr11* systems revealed the development of identical isoenzyme patterns during infection (Ludden and Daly, 1970). No great difference in chemical structure need be postulated; one might assume, for example, that in the *Sr6* system the protein surface buried during polymerization has one more amino acid residue than in the *Sr11* system, or that the protein is slightly more hydrophobic.

Gradations are clearly seen in oat stem rust resistance genes. To repeat information already given, the genes *Pg3* and *4* are effective only up to 20°C, *Pg8, 9, 13,* and *15* up to 25°C, and *Pg1* and *2* up to 30°C, the highest temperature tested. Where does one draw the line? If one had tested the genes at 20°C, one would have classed them altogether; if one had tested them at 23°C, one would have divided them, with the genes *Pg3* and *4* classed as temperature unstable and *Pg1, 2, 8, 9, 13,* and *15* as stable; if one had tested them at 27°C one would have classed *Pg8, 9, 13,* and *15* with *Pg3* and *4*.

Return to Table 6.5. The pathogen has an effect on the temperature response. To three cultures (126-6, 7; 21-2; 21-2, 3, 7) of the six avirulent cultures, Eureka wheat was fully resistant at 15°C; to the other three, the reaction of Eureka wheat was intermediate. Because it is unknown which cultures were homozygous and which heterozygous for avirulence, one cannot usefully probe further. There was a gradation from resistant (R), through intermediate (X), to susceptible (S). This gradation occurs commonly, as Johnson (1931) noted years ago. He found that any isolate of *Puccinia graminis tritici* that produced an intermediate X reaction on durum wheats at ordinary greenhouse temperatures was likely to produce a susceptible 4 reaction at higher and a resistant 0-1 reaction at lower temperatures. Similar behavior was demonstrated by Gordon (1930) for *P. graminis avenae,* and Peturson (1930) for *P. coronata avenae* when cultures that produced an X reaction were used. Other examples have been given for *P. recondita* (Johnson and Schafer, 1965) and *P. graminis tritici* (Luig and Watson, 1965; Luig and Rajaram, 1972). An intermediate X reaction type in wheat stem rust can be given by gene *Sr5* at 27°–30°C, by *Sr6* at 18°C, and by *Sr15* at 15°C, which are approximately the temperatures of change from effectiveness to ineffectiveness for the genes involved.

6.13.2 Second Temperature Test: The Effect of Temperature on the Dominance of Resistance

If temperature changes the dominance of resistance of the host, it changes it from dominant resistance at lower temperatures to recessive resistance at higher temperatures. So far as the data go, the test is successfully passed.

Luig and Watson (1965) and Luig and Rajaram (1972) studied the effect of temperature on the dominance of the resistance conditioned by the gene *Sr6* against *Puccinia graminis tritici*. They found a progression from dominant resistance at low temperatures to recessive resistance at higher temperatures, and then to ineffectiveness at still higher temperatures (see Table 6.8). A comparison of Tables 6.5 and 6.8 brings out the correlation, as temperature increases, in the sequence, resistance → X-type reaction → susceptibility, on the one hand, and the sequence, dominant resistance → recessive resistance → ineffectiveness, on the other.

In the literature of stripe (yellow) rust of wheat caused by *P. striiformis* there

TABLE 6.8

Effect of Temperature on the Expression of Resistance by Gene *Sr6* in Wheat against *Puccinia graminis tritici*[a]

Temperature (°C)	Culture 21-ANZ-2	Culture 126-ANZ-6 and NR7
15.6-18.3	Dominant	Dominant
18.3-21.1	Incompletely dominant	Dominant
21.1-23.9	Recessive	Incompletely dominant
23.9-26.7	Recessive	Recessive
26.7	Absent[b]	Absent[b]

[a] From data of Luig and Watson (1965).
[b] The gene *Sr6* was ineffective above 26.7°C.

are interesting fragments of the sequence, dominant resistance → recessive resistance → ineffectiveness. Table 6.6 shows, in the wheat variety Rego, the latter part of the sequence, from recessive resistance to ineffectiveness. But from the historical point of view special interest lies in the hint that the first part of this sequence can be found in the experiments of Biffen (1905). Early in the days of the revival of Mendelism, Biffen applied Mendel's laws to plant breeding for resistance, and was the first to do so. He crossed the resistant wheat variety Rivet with the susceptible variety Red King. The F_1 was susceptible; no relevant information was given that allows a guess about environmental temperature; all one can infer is that the plants were infected by the time when examinations stopped. The following year, 1903, was favorable for stripe rust, and rust appeared early, on March 16. The F_2 progeny were examined in the field. On June 29 there were 64 plants free from the disease and 194 infected, for the most part badly. From this 1:3 ratio he established that resistance was recessive, and accorded with Mendel's laws. But earlier, on June 15, when Biffen judged the epidemic to be at its climax, there were 78 plants free from disease, 118 plants mildly infected showing "a few pustules only," and 64 plants "badly diseased." This is roughly a 1:2:1 ratio. One infers that the heterozygotes remained resistant for 3 months after March 16, and that resistance changed from dominant to recessive only with the advent of summer.

Knott and Srivastava (1977) analyzed cultivars of common wheat showing good resistance to stem rust in the International Spring Wheat Nurseries. They isolated six hitherto unidentified genes for recessive resistance, and all of them conditioned only intermediate infection types 2+, 3. This should be read in conjunction with the evidence, given in the preceding section, that resistance giving intermediate infection types is commonly sensitive to temperature.

From the data of Cirulli and Ciccarese (1975) about the reaction of tomato plants to infection with various isolates of tobacco mosaic virus (see the previous section) it can be calculated that the gene *Tm1* conditioned dominant resistance to 53% of the isolates at 17°C but to none of them at 30°C; that the gene *Tm2* conditioned dominant resistance to 80% of the isolates at 17°C, but to only 34% at 30°C; and that the gene *Tm2a* conditioned dominant resistance to 58% of the isolates at 17°C, but to only 7% at 30°C.

Temperature manipulation could greatly help plant breeders to detect heterozygotes when resistance is normally recessive. For example, there is good reason to introduce stem rust resistance genes *Pg8* and *Pg9* into oats, because virulence in *P. graminis avenae* for these two genes tends to dissociate (see Chapter 2). But both these genes are recessive at ordinary temperatures, and the double task of introducing recessive genes would be daunting to any oat breeder. But if by the simple expedient of lowering the temperature it were possible to detect heterozygous resistance, the objection to recessivity would be resolved.

Genetical convention dictates that a capital letter be used to indicate dominance. Thus, the oat stem rust resistance genes *Pg1, Pg2, Pg3,* and *Pg4* are dominant and *pg8, pg9, pg11,* and *pg13* are recessive at ordinary temperatures. Because resistance is usually dominant at ordinary temperatures the custom has arisen of using capital letters to designate resistance and the lower case to designate susceptibility; thus *R* is for resistance and *r* for susceptibility. This is all very well when the temperature is appropriate, but confusing when it is not. The wheat stem rust alleles *sr6* for recessive resistance and, following custom, *Sr6* for dominant susceptibility at 24°C are, respectively, identical with *Sr6* for dominant resistance and *sr6* for recessive susceptibility at 18°C. When convention and custom clash, common sense must prevail. Throughout this book we use the custom of capital letters for resistance contrasted with susceptibility, and ignore the convention of capital letters for dominance contrasted with recessivity. Thus, we use *Pg8* and *Pg9* for the oat stem rust resistance genes, despite their recessivity, *Sr6* for the wheat stem rust resistance gene at any temperature, and so on.

6.13.3 Third Temperature Test: The Effect of Temperature on the Recessivity of Virulence

If temperature changes the recessivity of virulence of the pathogen, it should change it from recessive virulence at lower temperatures to dominant virulence at higher temperatures. This would make an easy test, but evidence seems to be lacking, one way or the other.

Dominant virulence in gene-for-gene disease has been reported in various pathogens. Of special interest is the connection in oat stem rust between infection type, temperature sensitivity of resistance, recessive resistance, and dominant virulence. The oat stem rust resistance genes *Pg8* and *9* condition an intermediate infection type 2 at 20°C (Martens *et al.*, 1979); they are sensitive to temperature, as we have already noted, becoming ineffective above 25°C; they are recessive at ordinary temperatures, as we have already noted; and virulence on them is dominant (Martens *et al.*, 1970).

Again we meet the phenomenon of variable infection types. Virulence in *Puccinia recondita* for the wheat leaf rust resistance genes *LrB* and *Lr14b* is dominant, and Samborski and Dyck (1976) remark that dominant virulence genes condition a reaction that is variable and difficult to classify.

The four-way relation, between infection type, the sensitivity of resistance to temperature, recessive resistance, and dominant virulence, goes to the core of the thermodynamics and energy relations of disease (irrespective of whether we want to relate it to a protein hypothesis, a polysaccharide hypothesis, or any other hypothesis). Special attention needs to be given to the effect of temperature on dominance/recessivity changes in host resistance and pathogen virulence. It is not necessarily to be expected that with increased temperature the rate of change from dominant to recessive resistance would be exactly matched by the rate of change from recessive to dominant virulence. The relative rates of change in a protein-for-protein hypothesis would depend on the relative contribution of host and pathogen to the protein polymer. Indeed, to view the matter in reverse, if one knew the relative rates of change, one should be able to calculate the relative contributions of host and pathogen to the protein polymer.

The four-way involvement of temperature just discussed would not be difficult to test experimentally, given the appropriate host lines (i.e., with homozygous and heterozygous resistance), the appropriate pathogen clones,and temperature-controlled environments. The experiment should stand high in the priority list. As we see it, there should be substantial temperature stability in all four ways, when resistance itself is temperature stable, that is, when no turning point in resistance exists between the cardinal minimum and maximum temperatures for disease; and there should be temperature instability in all four ways, when resistance itself is temperature unstable, that is, when there is a turning point between the cardinal minimum and maximum temperatures.

6.13.4 Fourth Temperature Test: The Effect of Temperature on the ABC–XYZ Systems

The ABC–XYZ system, discussed in Chapter 2, should increase in definition as temperature increases; that is to say, the association of virulences for resistance genes within the ABC group should increase with temperature, as should the association of virulences for resistance genes within the XYZ group, while the dissociation of virulences between these groups should also increase. This follows from the fact that virulence, not avirulence, is the controlling influence in grouping.

Evidence was given in Chapter 2. So far as it goes, it passes the test. This is the only temperature test that cannot be carried out in the laboratory or greenhouse. It demands field surveys; computerized surveys as suggested in Chapter 3 would soon settle the question.

6.13.5 Fifth Temperature Test: Three Levels of Enthalpy, Two Effects of Temperature

The discussion so far has implicitly involved two levels of enthalpy in proteins. (For a discussion of enthalpy, see Chapter 7.) The lower level is that of the unpolymerized monomers, i.e., of the protein subunits. The higher level is that of the protein polymer. This is just another way of saying that protein polymerization is endothermic. But there is a still higher level, associated with protein denaturation. A large literature exists of this, and an account has been given by Oosawa and Asakura (1975).

Denaturation is a breakdown of the internal structure of a protein, that can lead to a loss of polymerization, enzyme activity, or other functions of protein structure. Biological processes cannot continue after substantial protein denaturation. Permanent denaturation is therefore absent or at least not substantial at or near optimal temperatures for biological processes; it probably largely determines the temperature limit we call the cardinal maximum temperature of a biological process. Enzymes of thermophilic organisms have appropriate resistance to denaturation as, for example by having more salt bridges than their counterparts in organisms found in normal environments (Perutz, 1978), and by having a changed accessible surface area (Stellwagen and Wilgus, 1978).

From this we might expect that the polymers relevant to our discussion will begin to denature and dissociate as the temperature approaches the cardinal maximum. This would introduce resistance promoted by higher temperature, and existing only at or very near the cardinal maximum temperature. There is some evidence for this.

Rajaram *et al* (1971) found that the wheat varieties W3300 and W3303, which were susceptible to *Puccinia recondita* at 15°–18°C became resistant at 24°–27°C. Statler (1979) found that flax with the gene *L10* was susceptible to race 1 of *Melampsora lini* at 18°–24°C, but resistant at 29°–35°C. Gassner and Straib (1934) found that wheat susceptible to various races of *Puccinia striiformis* at 12.8°–17.0°C became resistant in greenhouses with a mean temperature of 23.0°C (ranging between means of 20.3 and 26.1°C) or 24.3°C (ranging between means of 20.0 and 27.2°C). To put these figures in perspective, the optimum temperature for *P. striiformis* is about 11°C, and the temperatures at which resistance occurred are approximately those at which spore germination ceases. To turn to bacterial disease, cotton plants (*Gossypium hirsutum*) with gene *b* and *BN* are susceptible to *Xanthomonas malvacearum* at night/day temperatures of 19°/36.5°C, but resistant at 27°–29°/36.5°C (Brinkerhoff and Presley, 1967).

There is thus evidence for a temperature effect that can be ascribed to denaturation. The effect is however unimportant in agricultural practice because the temperatures involved are well above the range at which natural epidemics occur, and resistance becomes available when it is no longer needed.

6.13.6 Shared Antigens of Host and Pathogen

If the proteins from host and pathogen produce a polymer larger than a dimer, they should share antigenic surfaces. Shared antigens of host and pathogen are known, and so far no explanation has been given for them other than in terms of protein polymerization. The experimental problem is, however, large. It derives from the fact that the relevant proteins are likely to be produced in quantity only after infection has been established, and by that time the analytical separation of the products of host and pathogen have become difficult.

How shared antigens are a feature of polymers can be illustrated in the simplest way by using head-to-tail polymers as an example. If in polymerization surface *a* is buried with surface *b,* i.e., if surfaces *a* and *b* are complementary, the subunits (monomers) of both host and pathogen are represented by (ab) and the polymer by (ab)(ab)(ab). . . . Surface *a* must be common to both host and pathogen protein, and so must surface *b*. (Otherwise polymerization would stop at the dimer.) And surfaces that are buried in polymerization are likely to be antigenically active, because (protein) antigen–antibody recognition in serology and protein–protein recognition during polymerization are based on the same reactions.

The work on serological cross-reactions in host and pathogen has been reviewed by DeVay and Adler (1976), among others. The evidence covers both wide and narrow specificities. Fedotova (1948) seems to have been the first in the field; she reported common antigens in the seed of cotton (*Gossypium hirsutum*) and *Xanthomonas malvacearum*. Her results were confirmed by Schnathorst and DeVay (1963) and DeVay *et al.* (1967), using leaves instead of seeds. If we confine ourselves to diseases listed in Table 6.1, Doubly *et al.* (1960) found that uredospores of the flax rust fungus *Melampsora lini* and flax plants had more antigens in common when compatible races of the fungus were used than when the races were incompatible. Golik *et al.* (1977) found antigens common to potatoes (*Solanum tuberosum*) and *Synchytrium endobioticum*. Palmerley and Callow (1978) found common antigens in *Phytophthora infestans* and susceptible varieties of potato. They also found common antigens between *P. infestans* and tomato and tobacco, but not between *P. infestans* and representatives of six higher plant families that are not hosts of *P. infestans*: mung bean, pea, radish, *Perilla, Pelargonium,* cucumber, and maize. Guseva and Gromova (1976) found common antigens in *Puccinia recondita* and wheat.

Attempts to find common antigens in other host–pathogen combinations, particularly the wheat–*Puccinia graminis tritici* combination, have failed. This is not unexpected. Because of analytical difficulties after infection has been established, it is usual to look for common antigens in ungerminated spores (or bacteria in axenic culture), on the one hand, and healthy plants, on the other. Because structural genes are normally repressed until they are needed, and they are needed only after infection has occurred, the customary procedure of separately analyzing healthy plants and spores (or axenically cultured bacteria) cannot be expected to detect all that goes on within the plant after it is diseased.

In their review of antigens common to host and pathogen, DeVay and Adler (1976) are reduced to the despairing suggestion that the selection of parasites by plant hosts, or vice versa, may be controlled by "fortuitous" homologies of their genomes. Almost any other suggestion would seem better than this. Argument by exclusion is dangerous; but we may at least note that the protein-for-protein hypothesis, published after DeVay and Adler wrote in 1976, is at present the only hypothesis that explains common antigens in plant disease and that it does so simply as a direct consequence of the infection process.

6.13.7 Divergent Mutation Rates to Virulence

The rate of mutation from avirulence to virulence varies greatly. In *Puccinia graminis tritici* avirulence for the wheat stem rust resistance gene *Sr15* easily mutates to virulence. But mutation from avirulence to virulence for gene *Sr26* is difficult; it has defied efforts with mutagens in the laboratory, nor is there evidence that it occurs in the field. These matters were discussed in Chapter 4. Greatly divergent mutation rates are relevant to the hypothesis because they point directly to proteins.

DNA has two complementary base pairs, adenine–thymine and guanine–cytosine; and a heritable change in the base sequence is defined as a mutation. From the simplicity of the arrangement great variation in mutation rates at gene level is neither expected nor found, provided that the amount and duration of radiation or other mutagen stays constant. But in the phenotype, as distinct from the genotype, the scope for variation in detectable mutation rate is immense, depending on how many mutations at gene level are passed on detectably to the phenotype. The protein-for-protein hypothesis shows neatly how this can occur; and the hypothesis accommodates divergent phenotypic mutation rates automatically.

Picture a relevant protein as consisting of 100 to 300, or more, amino acid residues of which about five to eight constitute the recognition system, i.e., they make up the surface buried during polymerization. (Proteins with as many as 1000 amino acid residues are known, and antigenic determinants are commonly surfaces of the order of five to eight amino acid residues.) If in the host a gene for susceptibility were replaced by one for resistance and this involved replacing one or more hydrophobic amino acid side chains by hydrophilic chains in the general body of the protein, a corresponding mutation from avirulence to virulence in the pathogen would be relatively easy. The pathogen's mutation could involve the replacement of one or more hydrophilic by hydrophobic side chains. Moreover, these replaced side chains would not have to be precisely positioned. But changes in the recognition surface would be quite another matter, and could involve impossible or difficult changes for the pathogen to match. For example, suppose that in the host the replacement of a gene for susceptibility by a gene for resistance involved in the recognition surface the replacement of an amino acid residue with a COOH, NH_2, or NH group that could make a salt bridge by a residue without one of these groups. The possibility of host surface–pathogen surface recognition by the salt bridge would be destroyed forever; the phenotypic mutation rate from avirulence to virulence in the pathogen would be zero, irrespective of the genotypic mutation rate. So, too, great differences between genotypic and phenotypic mutation rates from avirulence to virulence could be expected in the pathogen, if it had to match a host in which mutation from susceptibility to resistance involved the loss from the recognition surface of N, O, or S atoms that could form hydrogen bonds, or if it involved some form of steric hindrance to surface-to-surface contact.

It would not be profitable to pursue hypothetical examples of possible changes. But the fact remains that the protein polymerization hypothesis automatically introduces the concept of recognition surfaces; recognition surfaces would allow the host to make changes that the pathogen would find impossible or difficult to counter; and at protein level the experimental data about divergent mutation rates can be reasonably explained.

There is an important practical implication, even at the present time when techniques for studying the recognition surfaces are not available. If the hypothesis is correct resistance genes that the pathogen cannot match are more likely to be found in foreign species, i.e., in "nonhosts." (The gene *Sr26,* it will be remembered, was introduced into wheat from *Agropyron elongatum.*) Contrariwise, mutations to resistance in the natural host, e.g., from an *sr* to an *Sr,* and *lr* to an *Lr,* or a *yr* to a *Yr* gene in wheat, are likely to involve changes outside the recognition surface and to be easily matched by the pathogen's mutation.

Whereas a protein-for-protein hypothesis can, an RNA-for-RNA hypothesis cannot explain large differences in mutation rates from avirulence to virulence in the pathogen. This is another reason for believing that RNA is not involved in the gene-for-gene system except as messenger or in RNA–protein complexes.

6.13.8 Feeding the Pathogen

Evidence has been discussed in the preceding seven subsections about gene-for-gene recognition at the protein level. But the discussion must be extended beyond the possibility of mere recognition to the possibility that the gene-for-gene hypothesis is basically one of nutrition. There are several reasons for this extension.

1. All pathogens listed in Table 6.1 as gene-for-gene pathogens are biotrophic, either throughout the infection period or at least during the beginning of it (see Section 6.2). If the gene-for-gene system were only one of recognition, its association with biotrophy would be difficult to explain.
2. Specificity in gene-for-gene systems resides in susceptibility, not resistance (see Section 6.7). Feeding the parasite is the essence of susceptibility and parasitism.
3. There is a simple and obvious method of feeding, which will be discussed in due course.

Pathogens must feed. Their requirements for nitrogen are high. It has been calculated that in a badly rusted or mildewed field of wheat or barley the released spores carry the equivalent of a substantial application of nitrogenous fertilizer. All this nitrogen must cross the host/pathogen interface; and it must cross it efficiently, because in the rust fungi at any rate the amount of mycelium and haustoria is not large compared with the masses of spores that are released.

Plants that are wounded mobilize protein at the site of the wound; infection simulates or is in fact a wound, whichever way one cares to look at it; and there is microscopic and biochemical evidence that plants react to infection as they do to wounds. There is thus a very obvious way by which the pathogen can get host plants to deliver protein to it, provided that the pathogen remains biotrophic and keeps the host cells alive. Protein mobilization has the double merit, for the pathogen that it could supply it with food and for the plant pathologist that it could explain specificity in races by the million.

As background, note the obvious point that in sophisticated parasitism the host's biochemistry is reconditioned to suit the parasite. Viruses make the host treat virus RNA as mRNA. The crown gall organism *Agrobacterium tumefaciens* introduces some of its own DNA as a plasmid into the host's genome, to make the host synthesize the unusual amino acids octopine and nopaline which *A. tumefaciens* can utilize but the host cannot. The reconditioning of the host by rust fungi is to be seen in green islands, among other modifications. Making use of the host's wound-healing mechanism would seem to be relatively simple reconditioning. The pathogen's presence is a wound; the host throws its wound-healing proteins at it; and the pathogen has only to find some way of gathering the bounty.

There is a general similarity in protein synthesis in wounding and during infection. Peroxidases are the most studied wound proteins and infection proteins (Lipetz, 1970; Uritani, 1971); and something has already been said about them and about phenylalanine ammonia-lyase in Section 6.12. Chemical evidence is supported by the microscope.

In cereal rust infection the host cell's nucleus commonly moves toward the fungus haustorium, sometimes enveloping it. By analogy with nuclear movement after wounding, Gäumann (1946) referred to the movement of the host cell nucleus toward the haustorium as traumatotaxis. Gäumann was probably the first to connect infection with wounding, but did not follow the matter up. The connection is further illustrated by the term, wall lesion, that Hanchey and Wheeler (1971) use to describe the wall-like deposits or elaborations derived from the host cell and laid down on the host wall or around the invading parasitic cell. As Bracker and Littlefield (1973) point out, wall lesion is a good term that suggests a broader meaning than just wall deposits, the word lesion implying a localized abnormal structural change, injury, or impairment usually induced by some inciting factor. A wall lesion, as they define it, would include localized degradation or ruptures of the wall as well as abnormally induced swellings or growths or deposits on the wall.

In line with the protein-for-protein hypothesis we accept that a pathogen, be it a fungus, bacterium, nematode, or fly larva, makes its presence felt, and that the host cell responds as it would to other irritations, injuries or wounds.

More detail is given by Mendgen (1975) for compatible infections of *Phaseolus vulgaris* by *Uromyces phaseoli,* in an ultrastructural demonstration of peroxidase activity after the haustorium of the fungus had penetrated the host cell. All the Golgi bodies found around haustoria contained peroxidase activity. They secreted vesicles that embodied peroxidase, and "the vesicles were obviously secreted towards the partly invaginated sheath of the young haustorium." Around the haustorium and especially near the Golgi bodies, peroxidase activity was seen on the rough endoplasmic reticulum. Here we have visual evidence (at magnifications of 23,600 to 40,500) of the host cell delivering protein to the pathogen's doorstep.

From this and other evidence we accept that the site of the polymerization discussed in previous sections is at or very near the host–pathogen interface. Passage across the interface presents no physical difficulty; membranes are easily permeable to proteins, as shown generally by the excretion of extracellular enzymes. Across the interface the protein is presumably quickly digested. Unfortunately very little is known about lysosomes in fungi. Reiss (1972) has demonstrated in yeast an aminopeptidase that appeared to be compartmentalized at or very near the surface of the cell. Because protoplasts react to foreign proteins, and the host protein is foreign to the pathogen, we are probably safe in assuming that digestion occurs without deep penetration of the pathogen cell.

All this highlights the expected experimental difficulties in isolating the polymers chemically and demonstrating the *in vitro.* Apart from the difficulties mentioned in Section 6.13 there is the difficulty of impermanence. From formation to digestion the polymers probably move less than a fraction of a micron, and degradation probably follows synthesis almost immediately. A large accumulated reserve of host protein–pathogen protein polymer is not to be expected; indeed, if it occurred it would be contrary to the whole concept of the hypothesis.

On the question of peroxidase being the wound protein, or one of the wound proteins, there is evidence both for the abundance of isozymes and for their origin by mutation in duplicated loci, which would fit the fact of pseudoallelism of resistance genes (see Section 6.5). Hoyle (1977) demonstrated by isoelectric focusing as many as 40 different bands of isoperoxidase in commercial preparations of horseradish peroxidase. Since only a minority of isozymes are detectable, the real number is probably in line with the requirements of the protein-for-protein hypothesis. Scandalios (1974) studied 44 inbred lines of barley; and examined the genetics of four variant isozymes of peroxidase. Each of the variants was shown to be controlled by allelic genes at four distinct loci. He also studied maize. In maize there are ten electrophoretically detectable zones of peroxidase activity. Polymorphism at these ten regions is determined by allelic variation at each of ten distinct loci. Scandalios concludes that gene duplication followed by mutation at the duplicated loci is a likely mechanism which can lead to the type of isozyme multiplicity that he observed.

A hypothesis of wound protein as food for the pathogen is conceivable when the pathogen is a fungus, bacterium, nematode, or insect, but not when it is a virus. With a bacterial virus, the Qβ phage of *Escherichia coli,* Kondo *et al.* (1970) and Kamen (1970) have demonstrated a protein polymer, which is a tetramer with three subunits coded by the host and one by the virus genome. The protein is an RNA replicase. There has been a suggestion (Hariharasubramanian *et al.*, 1973; Hadidi and Fraenkel-Conrat, 1973), that a protein found in barley plants newly infected with brome mosaic virus is a polymer coded partly by the virus and partly by the host. It is possible that the protein-for-protein hypothesis is wider than its subsidiary offshoot of wound proteins would suggest.

6.13.9 Resistance Elicitors

According to the hypothesis of susceptibility discussed in the previous section, the pathogen injects proteins into the host cell, each protein being specified by an avirulence/virulence gene, and these proteins are matched and polymerized with proteins from the host. But when a protein specified by an avirulence gene is matched with a protein specified by a resistance gene, there is no polymerization. The pathogen protein is free in the host cell; and it is this free protein that, according to the hypothesis, elicits the defense reactions of resistance. Specificity of resistance is maintained by the pathogen protein. For example, suppose that an isolate of the wheat stem rust fungus avirulent for gene *Sr11* was tested on wheat with this gene. By hypothesis, there would be no polymerization with the protein specified by the host gene. Nevertheless, the avirulence in the isolate would be identified as avirulence for the resistance gene *Sr11* by the simple test that the isolate could attack (homozygous) wheat with the susceptibility gene *sr11,* other genes being equal, but not wheat with the resistance gene *Sr11*. Lines of wheat resistant and susceptible at the *Sr11* locus would, by the gene-for-gene relation, identify the avirulence gene in the pathogen.

This hypothesis, that the product of the avirulence gene is the elicitor, is supported by an experiment of Flor (1960). He induced two mutations from avirulence to virulence by X-rays in a uredospore culture of race 1 of *Melampsora lini*. There was evidence that at least one of these mutations was the result of a chromosomal deletion, and Flor explained the acquired virulence as being associated with the loss of the ability to form an inducer (elicitor). Virulence came about because the avirulence allele was made nonfunctional. In terms of our hypothesis, there was no longer any incompatibility between host and pathogen, because there was no longer any foreign protein loose in the host cell, the relevant structural gene in the pathogen having been deleted.

To summarize this and the previous section, both susceptibility and resistance to gene-for-gene disease are specific phenomena; but whereas susceptibility to gene-for-gene disease involves both a specific elicitor and a specific receptor, in resistance the receptor is unspecific, although the elicitor is specific.

6.14 SPECIFIC AND UNSPECIFIC RECEPTORS

The conclusion, that in gene-for-gene disease susceptibility is specific and involves a specific receptor and that resistance is also specific but involves an unspecific receptor, needs elaboration without special reference to a protein hypothesis.

To say that in susceptibility to gene-for-gene disease the receptor is specific is simply to paraphrase the gene-for-gene hypothesis. If the receptor were unspecific the disease could not possibly be on a gene-for-gene basis. In susceptibility therefore one can distinguish between two possible sorts of disease: disease with a specific receptor, as in gene-for-gene disease, and disease with an unspecific receptor, which is necessarily not a gene-for-gene disease.

As an illustration, consider *Agrobacterium tumefaciens,* the cause of crown gall. This pathogen introduces a tumer-inducing (Ti) plasmid into the host genome; it can induce and maintain tumors by introducing genetic information stably into plant cells. Consider biotypes 1 and 2 of the pathogen. They have different host ranges, and their Ti plasmids code for different metabolic and physiological properties. If it is found that the Ti plasmids of biotypes 1 and 2 have the same haven in the host genome, i.e., if they have the same receptor, crown gall is not likely to be a gene-for-gene disease; to make it a gene-for-gene disease would require supplementary hypotheses that are at present uncalled for. If they are found to have different havens, crown gall might conceivably be a gene-for-gene disease; at least, there is one degree of freedom in the argument.

Consider now resistance, not susceptibility. In resistance to gene-for-gene disease the receptor is (by hypothesis) unspecific, and this is possibly true for all resistance. We can widen the scope of the discussion by referring to "nonhosts." Müller (1950) studied the reaction of several angiosperms to infection by *Phytophthora infestans,* using a common race, presumably 0 or 4. Leaves of lettuce, dahlia, bean, and cabbage reacted like those of a resistant potato variety with the gene *R1*. The fungus penetrated the host and invaded the cells; the invaded cells started to collapse and darken; and the pathogen soon ceased to spread beyond the point of infection. Lettuce, dahlia, bean, and cabbage are nonhosts, and Müller found that they behaved like a resistant host. If it is true that we can class nonhosts with resistant hosts, it is reasonably certain that no specific receptor is involved, for two reasons. First, there are the numbers involved. There are thousands of pathogens each with hundreds of thousands of nonhosts (which is another way of saying that resistance is the rule and susceptibility the exception). To find enough specific receptors seems impossible. Second, specific receptors imply corresponding base sequences in parasite and host, and thus would imply coevolution of parasite and host. The probability is infinitesimally small that corresponding base sequences in the genomes of parasite and nonhost will be found, if parasite and nonhost have never been together. When in its evolutionary history did *P. infestans* ever evolve along with lettuce, dahlia, bean, or cabbage?

We conclude, then, that susceptibility to plant disease may or may not involve a specific receptor, according to whether the disease is a gene-for-gene disease or not; but that generally in all sorts of plant disease resistance probably does not involve specific receptors.

6.15 SACCHARIDES

We continue with the sequence, DNA→ RNA→ protein (enzyme) → saccharide, and must now consider saccharides.

Chemically, there is no reason to doubt the potential for polysaccharides to store variation on a large scale. There is a variety of sugars, variously linked, and with different anomeric forms. The number of possible permutations and combinations of sugar units in a polysaccharide is almost unlimited. Nevertheless we doubt that polysaccharides are involved in the storage and recognition of variation in gene-for-gene disease, for genetic reasons, discussed in this section, and thermodynamic reasons, discussed in the next chapter.

There is a detail in the sequence, DNA→ RNA→ protein→ saccharide relevant to variation. From the DNA in structural genes to protein, there is relatively little loss of variation (the loss being in samesense and nonsense mutations); but from protein to saccharide the loss is great. The variation in protein relevant to the sequence to saccharides is in enzymes; and it is in enzymes that a distinction is needed.

Isozymes are variants of an enzyme that have the same substrate and catalyze the same product. It follows that all qualitative isozyme variation is lost in the enzymatic production of saccharides. Isozyme variation, which is by far the most common known form of variation in enzymes, is relevant to a protein-for-protein hypothesis but irrelevant to a saccharide-for-saccharide hypothesis. To explain the gene-for-gene hypothesis one has to explain enormous variation (see Section 6.8); and one starts with a prejudice against involving polysaccharides when probably at least 99% of the variation in saccharide-coding genes is lost before saccharides are reached in the sequence. Whereas in the protein-for-protein hypothesis one could, and does, include isozymes and assume a single catalytically different enzyme, or at most a few, in a saccharide-for-saccharide hypothesis one must assume an abundance of distinct enzymes, each qualitatively different in catalysis.

Consider the matter further, in relation to particular enzymes and pseudoallelism. Roseman (1970) proposed a gene-for-gene relation in terms of a hypothesis of one gene–one glycosidic linkage. The heterosaccharides are built up, sugar by sugar, by glycosyltransferases. Each glycosyltransferase is specific for the transfer of a particular sugar, and each is coded by a single gene. We use glycosyltransferases to illustrate the variation problem. The *L* locus in flax has twelve pseudoalleles for resistance to rust. If host–pathogen recognition in the flax–flax rust system depends on saccharides, one must assume that the pseudoalleles code for twelve different glycosyltransferases each catalytically distinct. Because the pseudoalleles are likely to originate by tandem duplication, and because isozyme variation is the abundant form of enzyme variation, the probability of twelve separate, qualitatively distinct catalytic systems seems to be rather small.

Be this as it may, and despite skepticism about the amount of variation passed on to saccharide level, theories of saccharide involvement in gene-for-gene disease are current in the literature and must be treated on their merits.

6.15.1 The Suppressor Hypothesis

Glucans and various wall components are known to induce a hypersensitivity reaction. A 3-linked glucan has been found in the mycelial walls of *Phytophthora megasperma* var. *sojae* and in culture media in which this fungus has grown. Its biological activity is intense, as little as 10^{-13} moles being detectable when applied to soybean cotyledons. But Ebel *et al.* (1976) have found that this glucan is neither race-specific nor cultivar-specific. It neither reflects the difference between virulent and avirulent races of the pathogen nor the difference between resistant and susceptible varieties of the host. In *P. infestans* high molecular weight components of the fungus wall have been found to elicit a hypersensitivity reaction in potato tubers, but here too the response is unspecific; the literature is summarized by Doke *et al.* (1980).

To include unspecific elicitors in a theory of specificity in *P. infestans,* it has been suggested that there are suppressors of the hypersensitivity reaction which are specific (Doke and Tomiyama, 1977; Doke *et al.*, 1980). These suppressors are low molecular weight water-soluble glucans containing $\beta(1-3)$ linkages. They have been extracted from spores, mycelium, and from germination fluids (Doke, *et al.*, 1977, 1979, 1980). Suppressors from compatible *P. infestans* race 1, 2, 3, 4 suppressed hypersensitive cell death and the accumulation of the phytoalexin rishitin in tuber tissue of an *R1*-type potato variety that had been infected with incompatible race 4. A suppressor was also found in the incompatible race 4, but it was less active.

We reject the evidence that suppressors are responsible for specificity in gene-for-gene disease, for two reasons. First, the evidence misses the point, which is not to show a difference between compatible and incompatible races in their reaction to *R1*-type potatoes, but to show a difference between types *R1, R2, R3,* . . . or between virulence in *P. infestans* for *R1, R2, R3,* . . . The reasoning is clear from a comparison between Tables 6.3 and 6.4, and no elaboration is now needed. Second, the suppressor theory is contradicted by Flor's (1960) finding with flax rust that the deletion of an avirulence gene from the chromosome can turn avirulence into virulence.

6.15.2 Polysaccharides

Polysaccharide–polysaccharide complexes are well known, as between cellulose and xyloglucans (Keegstra *et al.*, 1973). Morris *et al.* (1977) found that extracellular polysaccharide (xanthan) of *Xanthomonas campestris* and *X. phaseoli* binds with typical galactomannan constituents of the plant cell wall. They suggest that this binding could lead to host–pathogen recognition. Because *X. malvacearum* is listed as a gene-for-gene parasite in Table 6.1, the suggestion must be noted. But Morris *et al.*, (1977) carried out no pathogenicity tests even at the level of Table 6.3, let alone that of Table 6.4. Moreover, they deduced (see their Fig. 12) that binding would be enhanced by reducing the temperature, whereas susceptibility of cotton to *X. malvacearum* is enhanced by increased temperature (see Section 6.13.1). We leave the matter there.

Agrobacterium tumefaciens has not been shown to be a gene-for-gene parasite; but experimental work on it bears on the topic of this section. For a summary of this work, see Sequeira (1978). Lippincott and Lippincott (1969) and Lippincott *et al.* (1977) showed that to induce tumors in pinto bean leaves the bacterium had to be attached to a specific wound site. Further, Lippincott and Lippincott (1977) suggested that the specificity of attachment involved the interaction between galacturonic acid residues in host cell wall pectins and a specific component of the bacterial envelope. This component may be a lipolysaccharide, and is specific in its attachment. However, the work they carried out does not approach the minimum test for gene-for-gene recognition illustrated by Table 6.4. It will be appreciated that this remark is not intended to dispute that there is some measure of host–parasite specificity; it is simply remarked that there is no evidence for specificity at the gene-for-gene level which is the topic of this chapter.

6.15.3 Lectins

Lectins are proteins or glycoproteins in which the activity resides in the protein moeity. They bind specifically with terminal sugars or several sugar units of an oligosaccharide. Specificity, then, is assumed to be protein-saccharide specificity.

A case for lectin-saccharide specificity has been presented around *Rhizobium* interactions, especially in the pairs *Glycine max–R. japonicum* and *Trifolium–R. trifolii*. Suggested details for these pairs vary somewhat.

In the soybean–*R. japonicum* pair the suggestion is that a saccharide component of the *Rhizobium* cell wall is specifically recognized by a lectin on the soybean root surface. Bohlool and Schmidt (1974) and, later, Bhuvaneswari *et al.* (1977) found that, with several exceptions, soybean lectin combined with strains of *R. japonicum* that could nodulate soybeans, but did not combine with strains that could not. On the other hand, Chen and Phillips (1976) found that soybean lectin bound to five of seven strains of rhizobia that do not nodulate soybean.

In the clover–*R. trifolii* pair the suggestion is that a lectin from clover roots binds with saccharides on the surface of both bacteria and root hairs (Dazzo and Hubbell, 1975; Dazzo *et al.*, 1976). The lectin thus provides a bridge between root hairs and bacteria. The difference between the suggestion for the clover–*R. trifolii* pair and that for the soybean–*R. japonicum* pair is that the former requires a specific saccharide on the root hair surface. Because of the doubt that in gene-for-gene systems specificity can reside in saccharides of the host, the difference might be important.

Looking at the evidence as a whole, one must echo the conclusion of Sequeira (1978), himself an authority on lectins, that the number of annoying exceptions is sufficient to question the validity of the lectin models. We do not here question the importance of attachment. But the attachment of bacteria to plant cell surfaces is only one step in the infection process, just as the attachment of a rust fungus to a leaf surface by means of an appressorium is only one step in its infection process. It is a feature of infection processes that there are recurrent specificities of varying fineness at different steps; thus, many (but not all) rust fungi penetrate their natural hosts better than they do nonhosts. Detection of a specificity, fine or coarse as the case may be, is no proof that it is *the* particular specificity which is responsible for characterizing gene-for-gene disease.

6.16 DISCUSSION

Gene-for-gene pathogens are diverse. The Hessian fly reaches its host on wings. Nematodes swim. Fungi arrive as spores borne in wind or water, or they swim. Bacteria splash on. Viruses have vectors, including man. Attachment and penetration are equally diverse. But gene-for-gene parasites have one common feature: They are all biotrophic, at least during the start of the infection process. They rely on the host for all their food, and the host must deliver the food to them, whether by diffusion or active transport. The pathogen, on its side, must keep the host cells alive.

The second feature common to gene-for-gene pathogens is the abundance of specific races (minimal specificity being defined in Table 6.4). Simple compounds, like sugars and amino acids, cannot explain this; and adequate qualitative diversity is likely to be found only in the polynucleotides, proteins, and polysaccharides. The later sections of this chapter were concerned in weighing the relative merits of these three classes of compounds, and the evidence, together with that in the next chapter, strongly points to proteins. Both recognition and feeding depend, the evidence suggests, on host protein–pathogen protein polymerization.

Direct chemical evidence *in vitro* of the hypothesis of polymers of mixed host–pathogen origin may be difficult to get, because the hypothesis would be best fitted by a transient polymer surviving only to move less than a micron from the host to the pathogen. Direct microscopic studies might be more fruitful. The sort of study carried out by Mendgen (1975) needs to be extended to the movement of peroxidase or other protein in the host cell at or near the surface appropriately defined and adjacent to the bacterium, fungus, nematode, or insect. Supplementary to this would be a study in fungi and bacteria of the distribution of lytic compartments (lysosomes) associated with protein degradation by the pathogen. Taking first things first, one might at least get more evidence about protein involvement, even if one did not get evidence for protein polymerization.

In biotrophy, i.e., in susceptibility, the host accepts the pathogen without reacting with processes seriously damaging to itself. Neither the pathogen nor its products are recognized as foreign bodies against which strong defense is needed. Resistance starts when the host recognizes and reacts to the pathogen or its products, and defends itself. In the protein-for-protein hypothesis it is a protein from the pathogen that is recognized in resistance. The reactions that follow are unspecific, and are as easily provoked by a glucan or copper chloride; but the protein itself can be traced to the avirulence gene that codes for it, and can thereby be identified. The unspecific reactions of resistance usually include the death of the host cell by hypersensitivity, though less commonly the pathogen or haustorium is sealed off. The death of the host cell presents the host with a second problem. A dead cell is large in terms of microbial dimensions, and without secondary transformation its protoplasm would be a good medium for secondary parasites, especially fungi. To cope with this second problem, which to the host plant is a wound problem, the host blocks secondary parasitism by the reactions we associate with hypersensitivity and, at least in some plant families and genera, by the release of phytoalexins. These reactions are postformed in relation to invasion by the gene-for-gene parasite, but preformed in relation to invasion by potential secondary invaders; and they are highly efficient as judged by the fact that secondary parasitism is normally scarce. Much literature has been devoted to whether the gene-for-gene parasite starts to die before the secondary reactions are initiated or after, or whether the death of the parasite starts the hypersensitivity and phytoalexin reactions, or vice versa. Sometimes, as with many cereal rusts and viruses, the parasite does not even die. These secondary reactions are not specific, and are not tied to a protein-for-protein hypothesis; they could equally be viewed in the context of any other explanation of gene-for-gene relations. The point is that resistance as defined by Flor's gene-for-gene hypothesis, specific as it is, must, if it leads to the death of cells, necessarily be followed by unspecific reactions to close the gates to secondary infection that dead cells inevitably are. The nature, speed, and timing of the gate closing are legitimate topics for research; but they must be seen as separate from the central biochemical topic of the gene-for-gene hypothesis, which is the molecular storage and recognition of variation.

6.17 OCKHAM'S RAZOR

Ockham's razor is the principle that hypotheses are not to be multiplied without necessity. (*Essentia non sunt multiplicanda praeter necessitatem.*) That in this chapter and the next a mass of extended detail is covered by a single hypothesis testifies to the benefit of the hypothesis.

This chapter, except in digressions about DNA, RNA, and saccharides, revolves around an hypothesis based on five premises. The first of these is implicit in the gene-for-gene hypothesis; the others are experimentally based. Susceptibility is specific, and (in the gene-for-gene hypothesis) requires both a specific elicitor (inducer) and a specific receptor; see Sections 6.7 and 6.14. Gene-for-gene parasitism is biotrophic, at least for a period after infection commences.

Great numbers of genes may be involved on both sides, involving massive storage and precise recognition of variation. When different recognition systems are involved, there is no allelism but pseudoallelism. Susceptibility in gene-for-gene disease is endothermic.

It would accord with Ockham's razor to wonder what range of disease could be included in the hypothesis. Is the hypothesis limited to a few diseases of the sort listed in Table 6.1, or does it extend widely? There is one indication that the hypothesis may cover a wide field: This indication is from the fact that in many diseases the pathogen establishes itself biotrophically even though it soon changes to necrotrophy. It would seem as if many pathogens which lack a food base are incapable of going over to necrotrophic attack until they have first established themselves biotrophically; and from this it is a short step to assuming that host–pathogen specificity may be established during the initial biotrophic phase, fleeting though this may be.

Colletotrichum lindemuthianum attacking *Phaseolus vulgaris* may be taken to represent not only anthracnose fungi but many others besides. In susceptible varieties (and we are discussing susceptibility) the hyphae are biotrophic for the first few days after infection. In the observations of Sinden (1937) the host plasmalemma is invaginated by a hypha and that invagination continues to the other side of the invaded protoplast so that the hypha becomes surrounded by host plasmalemma and cytoplasm. In the observations of Mercer *et al.*, (1974) invagination was not seen; instead, the plasmalemma was always displaced to one side, and the hypha grew between the cell wall and the protoplast before penetrating another cell. This relatively benign relationship between pathogen and host lasts only 4 to 5 days. It then ends abruptly; the pathogen degrades cell walls and kills protoplasts; and a spreading lesion is established. In resistant varieties the reactions are different, and the difference begins early. The decision whether the host–parasite interaction will be one of susceptibility or one of resistance is made during the biotrophic phase, and in this respect *C. lindemuthianum* is no different from, say, a rust fungus or *Phytophthora infestans*. Attempts to ascribe the host–parasite interactions that distinguish susceptible from resistant bean varieties to the necrotrophic phase, as when Anderson and Albersheim (1972) studied the endopolygalacturonases secreted by different races of *C. lindemuthianum* and the inhibitor proteins associated with the cell walls of different bean varieties, can be seen to miss the point that by the time necrotrophy begins, host–parasite interactions have already been decided.

Most plant pathogenic bacteria, other than those that cause soft rots, multiply in the intercellular space of their susceptible host plants for at least a brief period during the early stage of infection, and initially are not seen to injure the adjacent host cells. For a fleeting period they are biotrophic. In resistant host plants or in nonhost plants most pathogenic bacteria, other then those that cause soft rots, induce a hypersensitive reaction, just as *Phytophthora infestans* does in lettuce, dahlia, bean, and cabbage (see Section 6.14). Purely saprophytic bacteria fail to produce a hypersensitive reaction. This distinction between pathogenic bacteria in nonhost plants and bacteria which are not pathogens was first recorded by Klement and Goodman (1967). We have here a clue that pathogenic bacteria in their initial biotrophic establishment feed by producing what become elicitors of hypersensitivity if the invaded plant turns out to be a nonhost. The clue suggests the possibility that during the biotrophic phase parasitic plant bacteria behave much as, say, *Phytophthora infestans* does.

There are thus hints that a single hypothesis might cover host–pathogen specificity in all fungus and bacterial disease other than the soft rots, the tumor-forming diseases, and diseases caused by fungi that attack from an adequate food base. If this can be established experimentally, it would show Ockham's razor cutting cleanly.

Note Added in Proof

Browder (*Crop Sci.* **20,** 775–779, 1980) has listed 35 identified *Lr* genes for resistance to *Puccinia recondita*. The numerical foundation of the argument in Section 6.8 is evidently conservative.

7

Some Thermodynamic Background

7.1 INTRODUCTION

Thermodynamics gives evidence for involving proteins and against involving saccharides in gene-for-gene host–pathogen recognition. This short chapter began therefore as an appendix to Chapter 6, but it was decided that a wider discussion would be appropriate. Nevertheless the chapter's origin in proteins remains evident.

It is hoped that the chapter will bring into perspective some matters generally ignored in the literature of plant pathology, and will emphasize the importance of spontaneous reactions under thermodynamic control.

7.2 FREE ENERGY, ENTHALPY, TEMPERATURE, ENTROPY

The equation for Gibbs free energy is

$$\Delta F = \Delta H - T\,\Delta S$$

Here F (written G in Britain) is the Gibbs free energy, H is the enthalpy (= heat content), T is the absolute temperature, and S the entropy (= inherent probability). A spontaneous reaction always goes with loss of free energy. The equation shows that at constant temperature (i.e., isothermally) a loss of free energy can come about either by a loss of enthalpy or by a gain in entropy. Spontaneous reactions proceeding with a loss of enthalpy are called enthalpy-driven or simply enthalpic. They give out heat. They are exothermic. If spontaneous reactions proceed with the absorption of heat, i.e., endothermically, they necessarily involve an increase of entropy. They are entropy-driven or simply entropic.

Most spontaneous reactions are exothermic and give out heat. Endothermic reactions are relatively rare in biology. Lauffer (1975) lists some examples: the polymerization of protein, the division of fertilized eggs, protoplasmic streaming, the formation of pseudopodia in amoebae, karyokinesis. (Some of these are protein polymerization in disguise.) The proteins most widely studied in relation to polymerization are tobacco mosaic virus protein, collagen, myosin, actin, flagellin (in flagella), tubulin (in relation to microtubules), and normal and sickle-cell hemoglobin.

In protein polymerization the entropy increase comes mainly from the hydrophobic effect. When protein molecules come together in polymerization, bound water is released, and this release increases the entropy. It is the solvent that contributes largely to the entropy of the system as a whole. The term, hydrophobic effect, is misleading (Lauffer, 1975; Tanford, 1973). Lauffer uses the term, entropic union, instead. Nevertheless the hydrophobic effect, under this name, is commonly discussed. It derives from the presence at the surface of the protein molecule of amino acid residues with hydrophobic (''oily'') side chains. The amino acids concerned are tryptophan (the most hydrophobic of the essential amino acids), tyrosine, phenylalanine, methionine, proline, leucine, isoleucine, and valine. Alanine and glycine are neutral. The other essential amino acids have hydrophilic side chains, although arginine has a hydrophobic moiety within its hydrophilic side chain.

In protein polymerization there is also an enthalpy component contributed by hydrogen bonds between the buried surfaces. (These hydrogen bonds contribute substantially to the specificity of protein–protein recognition.) But this enthalpy component is overshadowed by the entropy component, and the net result is endothermic.

7.3 THERMODYNAMIC CLUES

By Le Chatelier's principle, endothermic reactions proceed further at higher temperatures, exothermic reactions at lower temperatures. The evidence in Section 6.13.1 shows, massively and consistently, that susceptibility in gene-for-gene disease is endothermic.

Protein–protein associations, as in protein polymerization, being endothermic, fit the evidence. RNA–RNA and polysaccharide–polysaccharide associations, established through hydrogen bonds, are enthalpic and therefore exothermic; the evidence excludes them.

Lipid–lipid or lipid–protein associations have not been considered. They are likely to be hydrophobic and endothermic, and therefore in the right direction. The reason for ignoring them is that they are normally unspecific, lipid associating with lipid indiscriminately to form mixed micelles. Without their having a precise recognition system, they do not qualify as the agents of variation and recognition in gene-for-gene systems. (Also, like polysaccharides, they suffer from not being able to reflect isozyme variation.)

Hybrid associations, RNA–protein or saccharide–protein, are not worth discussing until it is suggested what they are.

7.4 THE SOLVENT EFFECT

Protein polymerization is highly sensitive to the pH of the solvent. Lauffer (1975) records that tobacco mosaic virus protein in solution at 15°C and constant ionic strength of buffer was polymerized at pH 6.00 but unpolymerized at pH 6.25. When the protein from the Dahlemense strain of tobacco mosaic virus was used instead, polymerization occurred at temperatures above 10°C at pH 6.00, but at pH 6.25 did not occur at any temperature up to the highest, 20°C, tested. Other ions besides hydrogen ions actively affect protein polymerization. Among organic compounds in the aqueous solvent, sugars and other polyhydric compounds like glycerol promote protein polymerization.

The solvent effect introduces the possibility of modifying genes in gene-for-gene relations. Of special interest is the possibility of ontogenic effects. Mature-plant resistance is a wide term, applied to situations in which the young plant is susceptible and the mature plant resistant. At least some of it seems to be on a gene-for-gene basis. Thus, plants of the wheat variety Thatcher are susceptible to race 9 of *Puccinia recondita* when they are young but resistant when they are mature, whereas they are susceptible to race 161 both when young and mature (Bartos *et al,* 1969). It is not unreasonable to postulate solvent (sap) differences between young and mature plants. The problem is to identify which one, or which combination, of the the differences is relevant. A solvent effect which strongly promoted dissociation (depolymerization) of protein could in theory even induce a resistance reaction from *r* genes. There are possible examples of this in potato blight caused by *Phytophthora infestans*. The potato variety Record has no *R* genes, but its leaves of middle age react hypersensitively with *P. infestans,* whereas young and old leaves develop the normal, sporulating lesions characteristic of susceptible varieties with *r* genes (Populer, 1978). Similar results were reported by Warren *et al.* (1971) for the much more susceptible variety King Edward. This ontogenic effect may link up with the hypothesis of Grainger (1956, 1957, 1959), examined also by Warren *et al.* (1973), that potato plants without *R* genes are most resistant in middle age because the amount of "spare" carbohydrate is then least. Their method of estimation, as the ratio of total carbohydrate to residual dry weight of the shoot, is not precise enough to implicate particular sugars, although Warren *et al.* (1973) did find that sugar (glucose and fructose) injections increased susceptibility. Other evidence for a possible resistance reaction from *r* genes comes from the work of Estrada and Guzman (1969) and Thurston (1971) who observed a hypersensitive reaction characteristic of *R* genes when highly-resistant *andigena* lines of potato without known *R* genes were inoculated with *P. infestans*.

In passing we may note a dangerous interpretation currently being attached to an artifact, although the topic is outside our immediate field. Isolated protoplasts are infected with viruses, and differences in behavior found between the isolated protoplasts and intact plant tissue are ascribed to the absence of an intercellular reaction. Cells run on enzymes, some monomeric, some polymeric. When isolated protoplasts are bathed in a hypertonic solution, of a pH not necessarily precisely that of the relevant part of the cell, and of somewhat arbitrary ionic strength, it seems rash to ascribe differences in behavior solely to the protoplast's being alone.

7.5 A POSSIBLE THERMODYNAMIC SINK

Plant cells have an active transport system for protein; it can be seen in action in Mendgen's (1975) photographs. Parasite cells can also actively transport protein. Is there a no-man's-land at the wall lesion which is outside the sway of the host cell on its side and of the parasite cell on its side? If there is, how do protein polymers cross it? A possible answer is, they would be driven across down a gradient of free energy, provided that the polymer was more tightly bound, i.e., if it had a greater equilibrium constant, in the parasite than in the host. By adjusting, say, the pH on its side of the interface to give tighter polymerization, the parasite could make the polymer move spontaneously. The movement needed is only a fraction of a micron, so a steep gradient could in theory easily be arranged.

Some relevant equations are

$$K = [\mathrm{Sn}] \,/\, [\mathrm{S}]^n$$
$$-\Delta F^0 = RT \ln K$$

where K is the equilibrium constant when n moles of protein subunit (monomer) S combine to produce the polymer Sn, ΔF^0 means the standard change of free energy under conditions of unit activity for all participating molecules, R is the gas constant, and T the absolute temperature.

A thermodynamic sink of this sort would explain a minor difficulty. Protein polymerization in the host cell is loose, on the evidence of Chapter 6, but tight polymerization would be to the advantage of the pathogen because it would make mutation in the host from susceptibility to resistance difficult. Apparently the advantage to the pathogen of retaining as much free energy as possible in the polymer, in order to be able to extract it easily from the host, might outweigh any disadvantage.

Tanford (1978) has discussed in an easily available journal the movement of molecules to equalize chemical potential. The chemical potential is an intensive attribute with much the same relation to free energy as temperature has to heat. Tanford's discussion is about the assembly of molecules when each molecule under thermodynamic control searches for its position of lowest chemical potential. The matter is relevant to thermodynamic sinks, and it has been possible to simplify the discussion here because it can be assumed that there is no change in their number as the polymers move from host to pathogen.

8

Continuously Variable Resistance to Disease

8.1 INTRODUCTION

This chapter makes a change in genetic background. In previous chapters, except in Section 5.3, the variation of resistance in the host and of pathogenicity in the parasite was essentially discontinuous. For example, in Chapter 2 the surveys of *Puccinia graminis* in the United States, Canada, and Australia were described in terms of host plants that were either resistant or susceptible, with only rare mention of intermediates, or in terms of isolates of the pathogen that were either avirulent or virulent, with only rare mention of intermediates. Now, in the present chapter, the discussion is about continuous variation. Resistance varies from its lowest to its highest values without breaks, and intermediate values, far from being rare, are now the most common. For the chapters that follow, this chapter is a necessary bridge.

Other changes accompany the change from discontinuous to continuous variation of resistance. Genetic dominance featured largely in earlier chapters. Resistance in the earlier chapters was usually dominant over susceptibility, and resistance genes designated with capital letters, like *Sr11*, were frequently discussed. In the present chapter genetic dominance is less important and capital letters in gene designations disappear. Epistasis (nonallelic interaction) was at the center of Chapter 2, but will now be largely ignored. The downgrading of dominance variance and epistatic interaction variance automatically upgrades additive variance, which is taken to be the main component of the sort of genetic variance with which this chapter deals. Except mostly in Chapter 5, the resistance discussed earlier was vertical; here it is mainly horizontal. Flor's gene-for-gene hypothesis underlay most of the earlier chapters; it probably has little relevance in this chapter. Considered very broadly, these changes reach back to the breeding system of the host plants. Disease resistance varying discontinuously has been widely used by wheat breeders, and disease resistance varying continuously by maize breeders. Wheat is an inbreeder, maize an outbreeder; and this is not wholly irrelevant to the choice of type of resistance.

The volume of literature is no guide to the relative importance of discontinuously variable and continuously variable resistance in agriculture. With discontinuous variation, resistance genes can usually be identified and named; Mendelian ratios and concepts can be stated; and publications abound. With continuous variation, the plant breeder is often happy to record no more than that he has succeeded in producing a better variety, and this can usually be done in a paragraph. We take an immense amount of continuously variable disease resistance for granted. In agriculture an adequate amount of this sort of resistance is ordinarily a necessity, adequacy being judged in terms of environment and sources of inoculum.

Traits subject to continuous phenotypic variation are usually measured rather than counted, for which reason they are often referred to as metric traits, and the variation as quantitative variation. This chapter could be described as dealing with quantitative resistance, where previous chapters dealt mainly with qualitative resistance.

In this chapter we use the clumsy term, continuously variable resistance, for the sake of precision. Ordinarily the term, quantitative resistance, is adequate. The term, minor-gene resistance, is commonly used but is unacceptable, because, as will soon be shown, continuously variable resistance can be conditioned by only a few genes which are therefore necessarily major genes in effect.

As an example of continuously variable resistance, Hoff and McDonald's (1980) work on resistance to *Cronartium ribicola* in *Pinus monticola* may be cited. Resistance is manifested as a reduced frequency of spots on the pine needles, and Hoff and McDonald reported results of a pine-breeding experiment which are summarized in Table 8.1. The progeny obtained by crossing four pollen parents with several seed parents are classified according to the number of infection spots per linear meter of needle after inoculation with spores from infected *Ribes*. The distribution of seedlings in the progeny is roughly normal but slightly skewed, with seedlings most abundant at the intermediate level of 7 to 8 spots per meter of needle. For convenience the number of phenotype classes in

TABLE 8.1

The Frequency Distribution *Pinus monticola* Seedlings in Relation to Infection by *Cronartium ribicola*[a]

Number of needle spots[b]	Number of seedlings
0	0
1 or 2	4
3 or 4	9
5 or 6	18
7 or 8	23
9 or 10	12
11 or 12	7
13 or 14	6
15	1

[a] Data of Hoff and McDonald (1980). The data are for their two tests combined.
[b] The number of needle spots per linear meter of needle.

Table 8.1 was reduced to nine. But it would have been possible to extend the number indefinitely by recording the number of spots in needle lengths greater than a meter; and to say that the number of phenotype classes could have been extended indefinitely is to say that the variation is continuous.

With a few exceptions like that of blister rust of pines and stripe rust of wheat (Section 8.10), this chapter is written around maize diseases as examples. As has already been noted, continuously variable resistance has been particularly important to maize breeders, and keeping to one set of examples makes for both continuity and brevity of discussion.

8.2 THE POLYGENE MODEL

In most works, in genetics as well as in plant pathology, continuous variation and polygenic inheritance are treated as synonymous. In the early days of Mendelism, the term multiple factors was used to describe the genes thought to be involved in continuous variation, because it was thought that continuously distributed characters depend on genetic differences at many loci. Mather (1943) introduced the adjective, polygenic, into the literature. It was assumed that in the polygenic inheritance of a character, independent segregation occurs at an indefinitely large number of loci affecting the character, and that the effects of allelic substitution at each locus are trivial compared with the total amount of observed variation in the character. There are other assumptions which we need not discuss because they barely enter our story.

The segregations with increasing numbers of independent allelic pairs follow the expansion of the binomial $(r + s)^n$, where n is the number of segregating alleles, and r and s are alleles of resistance and susceptibility, respectively. The symbols R and r for resistance and susceptibility are out of place and avoided, because there is no implication of dominance. In a one-gene model, with $n = 2$,

$$(r + s)^2 = r^2 + 2rs + s^2$$

This expansion can be seen to represent an F_2 segregation of 1:2:1, with the population in three genotype classes. In a corresponding two-locus model the expansion is

$$(r + s)^4 = r^4 + 4r^3s + 6r^2s^2 + 4rs^3 + s^4$$

This can be taken to represent a segregation of 1:4:6:4:1, with the population in five genotype classes. With increasing number of segregating loci, the binomial expansion produces a frequency pyramid of F_2 segregations:

1 locus	1:2:1
2 loci	1:4:6:4:1
3 loci	1:6:15:20:15:6:1
4 loci	1:8:28.56:70:56:28:8:1

The number of genotype classes increases as $2n + 1$. When n is large enough, the number of classes will be large enough to give a continuous distribution, without distinguishable intervals between classes. The binomial expansion for an F_2 population with an infinite number of loci produces a normal probability distribution; and the creation of a large number of genotypic classes is the basis of the polygene model of continuous variation. The F_2 population is introduced into the discussion simply for convenience. The essential point is that with many loci segregating independently a natural population is capable of division into many classes, with the result that variation is for all practical purposes continuous. Two points should be noticed about the model.

First, the loci are considered to be equal in effect, so that none of them has a large effect. The polygene model prescribes not only that many genes should be involved, but also that none of them should have a large effect.

Second, the variance was taken to be additive. Complete dominance of resistance would change the 1:2:1 segregation to a 3:1 segregation, with two phenotype classes instead of three, the 1:4:6:4:1 segregation to a 15:1 segregation, with two phenotype classes instead of five, and so on. The variation would remain markedly discontinuous, however many loci were involved.

8.3 FOUR OTHER MODELS

The polygene model has never been proved experimentally; it remains a surmise. What is most unfortunate is the usual treatment of continuous variation and polygenic inheritance as synonymous. That polygenic inheritance will produce continuous variations is indisputable. It is the converse, that continuous variation implies polygenic inheritance, that is wrong. We describe now four models of continuous variation that is not polygenic in origin.

8.3.1 The Oligogene Model

The polygene model of continuous variation postulates that all the loci have equal (and small) effects. When once the postulate of equal effects is discarded, continuous variation can be based on oligogenes. This was first pointed out by Thompson (1975). Continuous normal phenotypic variation can be determined by the segregation of only three genes, one of these having three times, and one twice, the effect of the third gene. The distribution can be approximately normal, even though more than 90% of the variance is contributed by only two loci.

One should perhaps call this the unequal-effect model rather than the oligogene model, because its essence is the inequality of effect.

8.3.2 The Linked-Loci Model

A variant of the previous model allows the postulate of equal effects of genes to be retained, by postulating unequal effects of gene blocks. Instead of postulating a gene with three times the effect of another gene, one postulates a block of three linked genes. Genes in blocks have theoretical advantages, especially in a biochemical background where increased catalytic activity could arise from the tandem duplication of structural genes.

8.3.3 The Monogene Model

Environmental variance has not yet been discussed. It is often large and can account for the bulk of the phenotypic variance. When this happens, even monogenic resistance without complete genetic dominance and with segregation approximately 1:2:1 can underlie continuous variation. Resistance to cucumber mosaic was for many years considered to be polygenic, until tests were made under more strictly controlled environmental conditions when it was found to be monogenic (Walker, 1966). Hoff and McDonald (1980) found that their data, summarized in Table 8.1, on resistance in pine to *Cronartium ribicola* were best fitted by assuming monogenic inheritance. Heritability was low, being estimated at 0.34 for one tester group and 0.40 for the other.

One should perhaps call this the model of high nonheritable variance, rather than the monogenic model, because nonheritable variance is where the continuous distribution comes from. But the concern of this section is to relegate the polygene model to its proper place, and classifying models in terms of gene numbers makes for clarity.

8.3.4 The Composite Model

Four models have been discussed with clear differences between them. Reality probably does not lie wholly in any one of them, but in a mixture. Continuous phenotypic variation of resistance is probably determined in most instances by a large number of genes of small effect and a few of large effect, and by environmental variance that is sometimes large. Linkage may be important. It is the few genes of large effect that make resistance oligogenic or monogenic for all practical purposes of plant breeding.

Consider first the genes of small effect by means of an example. Tewari and Skoropad (1976) studied the relationship between epicuticular wax and blackspot caused by *Alternaria brassicae* in rapeseed. Free water on the host's surface promotes infection, which is precluded if water droplets fail to remain on the leaf. It was shown that the numbers of conidia and the subsequent development of blackspot lesions were significantly greater per unit area of leaves that lacked a waxy bloom. Clearly, any genes involved in forming a waxy bloom are genes for resistance. With disease considered in general, genes of the sort that relevantly change the cuticle or the stomata or any part of the host that is infected or genes that control the growth form of the plant in a way that affects the microclimate or genes that modify the host plant biochemically or physiologically are all potentially genes that hinder or promote infection and are therefore genes for resistance or susceptibility. Some may well have a large effect, but there must be many that make only a trivial or very small contribution to the total genetic resistance.

Almost all the analyses that have been made of real situations reveal that segregation at only a few loci contributes most of the genetic variance in the quantitative inheritance of resistance. It is with them that practical plant breeding is mainly concerned. Lest terminology be misunderstood, it will be repeated at the risk of being tedious that the involvement of a large number of genes does not necessarily make resistance polygenic in the accepted meaning of the word. Polygenic resistance is resistance conditioned by many genes segregating independently and with the qualification that none of the genes has a large effect. The reason for the qualification is easily seen from an example. If 100 genes conditioned resistance, with one of them contributing 95% of the genetic variance, the resistance would be shown by genetic analysis to be monogenic. Even if the 5% of the variance contributed by segregation at the other 99 loci were not swamped by nonheritable variance, the contribution of each of these loci individually would be undetectable.

8.4. POLYGENIC RESISTANCE VERSUS BREEDING FOR RESISTANCE

For many years theory and practice about polygenic resistance have been separate. In theory, at least until Thompson's (1975) paper appeared, continuous variation of resistance was taken to be polygenic. In practice plant breeders carried on as if continuously variable resistance was monogenic or oligogenic; and in essence they were usually correct. If disease resistance were truly polygenic, the prospects of breeding new resistant cultivars would be difficult.

Linkage is one problem. Consider an illustration of what commonly occurs in practice. There are two varieties of the host. One of them is agronomically excellent in the absence of disease, but is very susceptible. The other is highly resistant to disease, but agronomically poor. It is desired to combine the agronomic excellence of the one with the high resistance of the other. They are crossed, and in segregating progeny of different generations the plants with most resistance are selected. The problem is to get agronomic suitability along with the resistance. If the resistance is monogenic or digenic, it can be introduced with relatively little of the genetic material of the agronomically poor ancestor, and with luck the worst of the linkages could be broken. But if the resistance is polygenic in the true sense of the word, with many genes all segregating independently, it will remain linked with agronomic poorness. Selection for resistance will be selection for agronomic unsuitability; selection for agronomic suitability will be selection for susceptibility to the disease.

There are other disadvantages of polygenic inheritance besides linkage. Breeding for resistance is largely a matter of keeping the resistance intact, i.e., of not dispersing the resistance genes. This is particularly important when resistance is introduced by backcrossing. If resistance were polygenic, dispersion would be inevitable and plant breeding correspondingly difficult.

Scrutiny of the literature has revealed no instance of successful breeding for resistance that involved transferring very many genes from agronomically poor sources. There are claims by plant breeders that polygenic resistance has been successfully used, but the claims are for quantitative, continuously variable resistance that is probably oligogenic but mistakenly assumed to be polygenic. Very often, discussions of "polygenic" resistance are coupled with the statement that resistance is conditioned by few genes (see Section 8.6).

8.5 THE ERROR OF EXPECTING SAFETY IN NUMBERS

It was commonly argued that polygenic resistance was likely to be stable. Safety, it was reasoned, lay in numbers. The more resistance genes the pathogen had to overcome, the less likely it was to overcome them. Monogenic resistance, the reasoning ran, was unstable, because the pathogen could easily mutate to match the resistance by added virulence; polygenic resistance would require a multitude of matching virulences that the pathogen could not accomplish and the resistance would be stable. The reasoning has become largely hypothetical, because it is now reasonably certain that polygenic resistance does not commonly occur. But even if it did occur, the argument about there being safety in number is sound only within limits.

It is probably safe enough to argue that the more resistance genes there are, the better the resistance will be, provided that the genes are all of the same kind. It is when genes are of different kinds that the argument about safety in numbers falls away. In Canada wheat cultivars or multilines with ten *Sr* genes, *Sr5, 7a, 7b, 8, 9d, 9e, 10, 11, 14,* and *Tt1,* would be susceptible to stem rust, because most isolates of *Puccinia graminis tritici* in Canada are virulent for all of them. But maize with the single gene *Ht1* would be adequately resistant to *Helminthosporium turcicum*. The point is, the *Sr* genes and the *Ht* gene are different, and there is no reason to expect that ten of the one kind are necessarily better than one of the other. Differences in number do not necessarily compensate for differences in kind. But with resistance of the same kind, the more genes there are, the better the resistance is likely to be. The homozygote *Ht1Ht1* conditions better resistance than the heterozygote *Ht1ht1,* and the two genes *Ht1* and *Ht2* condition better resistance than either of them singly, better resistance being expressed as smaller lesions with less necrosis (Hooker, 1979).

The suggestion in the preceding paragraph, that the more resistance genes of the same kind there are, the better the resistance is likely to be, must be qualified to exclude genes all of an ABC group or all of an XYZ group. This is clear enough from the data in Chapter 2.

The fact that ten *Sr* genes do not make wheat safe from stem rust whereas one *Ht* gene normally makes maize safe from northern leaf blight, destroying as it does faith in mere numbers, emphasizes the need always to bear in mind the sort of resistance with which one is dealing.

8.6 EXPERIMENTAL EVIDENCE ABOUT GENE NUMBERS

Evidence has already been given that resistance in pine to *Cronartium ribicola* and in cucumber to cucumber mosaic virus is conditioned mostly by one gene. What about other continuously variable resistance? There is much indirect evidence, based on the quick accumulation of resistance by mass selection and by the success in obtaining resistant maize inbreds by backcrossing, that gene numbers are small. But direct evidence is scarce, because of experimental difficulties.

Hughes and Hooker (1971), from their own experiments and those of Jenkins *et al.* (1954), concluded that resistance in maize to leaf blight caused by *Helminthosporium turcicum* is inherited simply and conditioned by few genes; they suggested from three to six. Grogan and Rosenkranz (1968) could not satisfactorily determine the number of gene loci controlling resistance of maize to maize stunt virus; the indications were that the inheritance was simple, but with more than one gene pair involved. Kim and Brewbaker (1977) used two maize inbreds highly resistant to rust caused by *Puccinia sorghi* and crossed them with susceptible inbreds. They estimated that one of the resistant inbreds Oh545 had two gene pairs conditioning resistance. Estimates for the other resistant inbred Cm105 by three different methods averaged 1.3 gene pairs, which presumably indicates one pair of large effect and others of less effect. Parlevliet (1976) studied the period of latency of infection by *Puccinia hordei* in four cultivars of barley. (The period of latency is the interval between inoculation and resulting sporulation; a long period is a component of high resistance.) He estimated that the period was controlled by one gene of large effect and four or five of small effect. In other experiments Parlevliet and Kuiper (1977) estimated the number of genes to be more, but not many more, than three.

To get these estimates in perspective, one might remember that the wheat cultivars Selkirk and Canthatch each have at least five *Sr* genes for resistance to stem rust (see Section 2.17). There is nothing to suggest that this is unusually many for wheat cultivars, and nobody would ever describe the resistance given by *Sr* genes as polygenic. Yet the number of genes involved is, on the evidence, of the same order as that involved in the continuously variable resistance discussed in the previous paragraph and commonly called polygenic. In the literature of resistance to plant disease the prefixes oligo (or mono) and poly have little to do with numbers.

8.7 THE CENTRAL ROLE OF ADDITIVE VARIANCE

Gene number does not determine the continuous variation of resistance. What does? The tentative answer seems to be, additive variance. Additive variance is implied in all the models discussed in Sections 8.2 and 8.3: the polygene model, the oligogene model, the linked-loci model, the monogene model, and the composite model. It is a constant theme of the biometric analyses of experimental data. If in reading the literature you come across the words, "polygenic resistance" you can with reasonable safety delete them and substitute "genetic variance mostly additive."

Additive variance is what makes progeny resemble parents on both sides. If all the variance is additive, an F_1 hybrid will fall midway between its parents. This contrasts sharply with the dominance variance associated with the *Sr, Lr, Pg,* and the *R* genes discussed in earlier chapters; if dominance is complete, the hybrid takes after one parent and not the other.

The literature is consistent in associating continuously variable resistance with additive variance. Nonadditive variance may or may not also occur. Russell (1961), working with resistance to the stalk-rot in maize caused by *Diplodia maydis,* found that additive effects were of considerably greater importance than nonadditive effects. Kappelman and Thompson (1966), working with resistance to the same disease, found additive gene effects in all the maize populations they tested, together with dominance effects in six out of eight population. Hughes and Hooker (1971) studies the "lesion-number" form of resistance in maize to *Helminthosporium turcicum*. (This is distinct from the "chlorotic-lesion" resistance conditioned by the genes *Ht1* and *Ht2* discussed in Chapter 5. The lesion-number form of resistance also conditions lesion size, and is the form of resistance with a long history in maize breeding.) Additive gene effects were of major importance and were detected in all populations in both years of study. Nonadditive gene effects were variable, and their relative importance depended on the population involved and the year of study. In the nonadditive genetic variance there was evidence for both dominance and epistasis. Moll *et al.*, (1963), working with the inheritance of resistance to brown spot caused by *Physoderma maydis* in maize, concluded that much of the genetic variance was additive. Evidence for dominance was not found, but epistasis appeared to be present in some populations. Lim (1975) studied resistance in maize to race T of *Helminthosporium maydis,* and used both normal and *cms-T* cytoplasms. Disease resistance in eight parents, each duplicated for normal and *cms-T* cytoplasm, was predominantly additive with some partial dominance. Epistasis was unimportant or absent. Kim and Brewbaker (1977) found both additive and nonadditive variance in the resistance of maize to *Puccinia sorghi*. Nonadditive variance appeared to make a major contribution only in the inheritance of rust resistance among progenies of the moderately resistant inbreds Mo17 and C121E. Resistance in maize to maize stunt virus seems to be all additive (Grogen and Rosenkranz, 1968).

8.8 ADDITIVE VARIANCE AND STABLE RESISTANCE

Resistance called polygenic has had a high reputation for being stable. Now that we discard the general concept of polygenic resistance and replace it with concept of (mainly) additive resistance, additive resistance must inherit the reputation for stability. On what evidence was the reputation based?

Discontinuously variable resistance has been used in maize. Examples are the resistance given by the genes *Ht1* and *Ht2* against *Helminthosporium turcicum,* by normal cytoplasm against *H. maydis,* by the recessive gene *rhm* against the same fungus (Smith and Hooker, 1973; Smith, 1975), and by a dominant gene to race 1 of *H. carbonum* (Ullstrup and Brunson, 1947). Nevertheless it is on continuously variable resistance that maize breeders have mainly relied, and this resistance has proved to be very stable. Continuously variable resistance to *Helminthosporium turcicum* and *Puccinia sorghi* illustrate the success maize breeders have achieved.

Not much was heard about *H. turcicum* in the old days of open-pollinated maize. But when hybrid maize was introduced many of the early inbreds were very susceptible; during the process of inbreeding not enough attention was given to disease resistance, and many loci became homozygous for susceptibility where previously open-pollinated crops were heterozygous and heterogeneous. In 1942, at the time when hybrid maize was largely replacing the old open-pollinated varieties, there was a destructive epidemic. During 1942, Elliot and Jenkins (1946) tested 38 commercial inbred lines of corn-belt maturity at Beltsville, exposing them to natural conditions of infection. In 1944 they repeated their tests, in this year using artificial inoculation. They used a scale of rating the amount of disease. In 1944 by August 25 half the inbred lines had reached the highest disease rating ("Very heavy infection, lesions abundant on all leaves, plants may be prematurely killed."). In 1942, without artificial inoculation, 15 out of the 38 inbred lines had by August 12 reached the stage of abundant lesions on the lower leaves. Only 12 of the lines had developed no more than slight infection by August 12. The general high level of susceptibility of the new inbred lines of corn-belt maturity explained the 1942 epidemic. (Late-maturing inbred lines fared a little but not much better.) The dangerous situation was not allowed to last long. Jenkins *et al.* (1954) demonstrated that genes for resistance could be quickly and adequately concentrated in two cycles of recurrent selection, and inbreds with appropriate resistance were soon used. The resistance has persisted, despite the fact that the fungus is variable, with different forms pathogenic to maize, Sudan grass, sorghum, and Johnson grass and with large differences in aggressiveness between cultures. This pathogen variability has not resulted in the loss of resistance to *H. turcicum* leaf blight in agriculture (Hooker, 1979), and there is no evidence to suggest that the continuously variable resistance accumulated by maize breeders since 1942 can be negated by pathogenic variation.

Discontinuously variable resistance in maize to *Puccinia sorghi* is well known. It is conditioned by *Rp* genes, comparable with the *Sr, Lr,* and other genes in wheat and oats discussed in Chapter 2. But very little use of it has been made by maize breeders.

Puccinia sorghi is potentially destructive under moist conditions. Losses in grain yield of 6.3 to 23.5% have been measured in susceptible hybrids (Hooker, 1962b; Russell, 1965). However, resistance has kept loss of yield low. An adequate level of continuously variable resistance exists within the germ-plasm pool of the American corn belt (Hooker, 1979). When inbred lines were first developed in the United States at the beginning of hybrid maize production, maize breeders had many of these lines available and they selected heavily against susceptibility for rust. The resistant lines which were selected provided in due course the germ plasm base for the newer inbred lines that now make up many of the modern hybrids. Consequently, high susceptibility to *P. sorghi* is rare among inbred lines in the United States (Hooker, 1979), even though many maize breeders pay little attention to rust. There are indeed some inbred lines which are susceptible, but because of heterotic patterns of grain yield these are in the normal course of events crossed with more resistant lines, and the resulting hybrids have adequate resistance.

If one considers hybrid lines as distinct from inbred lines, there is enough resistance to rust to keep yield loss under control, this resistance is continuously variable, and resistance has been maintained over the years. Evidence for the development of races of *P. sorghi* that could circumvent this resistance is absent.

Resistance has remained stable despite the cultivation of maize over immense and continuous areas, and despite the variability of *P. sorghi* aided by a sexual phase on *Oxalis* spp. which are often heavily infected in Mexico (Borlaug, 1946). Uredospores of *Puccinia* are notorious for their ability to travel, and the fact that there have been no reports of having to change maize hybrids because of the appearance of new races of *P. sorghi* suggests that the continuously variable resistance is horizontal.

What has been said for *Helminthosporium* leaf blight and rust hold in general for continuously variable resistance to disease in maize. The current level of resistance of North American maize hybrids is quite effective in disease control. As Hooker (1979) points out, the 1975 season was wet and should have favored many leaf diseases. In fact, few disease problems were encountered, and grain yields were the highest on record. Much improvement is still possible. That improvement would be to raise the level of resistance even further; it would not have to do with repairing lapses of resistance, because this, it seems, is unnecessary.

8.9 ADDITIVE RESISTANCE IN GAINS BY SELECTION

In relation to plant breeding for resistance to disease polygenic inheritance, on the one hand, and additive variance in a system of oligogenic inheritance, on the other, are poles apart. Polygenic inheritance would be a burden on the plant breeder. The larger the number of gene pairs involved, the smaller is the chance of getting a phenotype at a given distance from the mean phenotype; getting a resistant phenotype at a distance from the mean phenotype is precisely the breeder's aim. In contrast to this, additive variance in continuously variable disease is the form of variance most amenable to manipulation by plant breeders.

Additive resistance is more amenable to change under selection pressure if only a few gene pairs are involved; this is the simple converse of what was said about polygenes in the previous paragraph.

Additive resistance is also more amenable to change through selection if nonadditive effects are absent. Commonly, in the course of selection for resistance one chooses as parents individuals that show the most resistance. With purely additive variance the first offspring generation would have the same mean resistance as the parents had. Any average gain of resistance through selection in later generations would be passed on in full to the next generation. Dominance retards change under selection, because dominance causes individuals phenotypically extreme to have progeny on the average less extreme. This is known as regression. Any average gain of resistance through selection would not be passed on in full to the next generation. There would be some slipping back toward the mean before selection. When with continuously variable resistance dominance is present one needs more cycles of selection to obtain a gain of given magnitude than when it is absent. Epistasis, which is dominance between nonalleles, also causes regression.

In practice, gains of resistance to disease through selection have been rapid. Some examples from maize breeding will suffice. Jenkins *et al.* (1954) increased resistance to leaf blight caused by *Helminthosporium turcicum*. They started with nine populations each of about 250 plants, and used a technique of recurrent selection after inoculating young seedlings. By flowering time symptoms were clear. The ten most resistant plants in each population were selected at flowering time, their pollen was collected, mixed, and placed on the silks of the same ten plants. The seed obtained in this way was used to start the next cycle of selection. Response was rapid. In each of the nine populations two cycles of selection sufficed to raise resistance to a level which made loss from disease negligible in agricultural practice. *Ht* genes for resistance were not involved in this work.

The genetic advance under selection has been equally swift with maize rust caused by *Puccinia sorghi*. Kim and Brewbaker (1977) worked with this disease in Hawaii in the field under conditions of considerable infection. In crosses between moderately resistant and susceptible inbred lines, they obtained lines in the F_2 population that were as resistant as the more resistant parental lines. To accumulate resistance they selected the most resistant 5% of the plants, and when the three most resistant inbreds had been used in the crosses the genetic advance averaged 1.73 units per generation on a scale 1 = resistant and 7 = very susceptible.

Miles *et al.* (1980) studied resistance in maize to leaf blight caused by *Helminthosporium turcicum,* stalk rot caused by *Diplodia maydis,* and leaf blight and stalk rot caused by *Colletotrichum graminicola*. They reported that quantitative disease resistance responded readily to recurrent selection, and that selection for disease resistance with artificial inoculation could in two to three cycles produce resistant populations for subsequent yield selection. The swift gain of resistance by selection is itself evidence that resistance is not polygenic in the true meaning of the word, and plant breeders have reason to be thankful for resistance that is both oligogenic and largely additive.

8.10 TRANSGRESSIVE SEGREGATION AND POLYGENIC RESISTANCE

Transgressive variation allows individuals to appear in the F_2 or later generations which show a more extreme development of a character than either parent. With clonally propagated crops, with heterozygous parents, transgressive variation is apparent in the F_1 generation as well. It is generally accepted that transgressive variation arises from the cumulative and complementary effects of genes originally present in the parents of the hybrid.

Transgressive variation has long been used in breeding for resistance to disease. For example, the accumulation of resistance to *Phytophthora infestans* in potatoes in the decades following the great epidemics of the 1840s in Europe was the result of selection of transgressive variants. But the use of transgressive variation in breeding for resistance has now been given a new impetus in attempts to avoid vertical resistance. Several writers, notably Robinson (1979),* have recommended that parents be selected which are not highly resistant, but Krupinsky and Sharp (1979) seem to have been the first to spell out the genetic aims in terms of transgressive segregation and polygenes.

Krupinsky and Sharp worked with stripe (yellow) rust of wheat caused by *Puccinia striiformis*. They set out to determine whether adequate continuously variable resistance could be accumulated if resistant parents or resistant phenotypes in the F_1 or F_2 generation were excluded. They wished to avoid resistant progeny in early generations in order to try to avoid genes tainted with dominant vertical resistance; with resistant parents, selection for resistance in early generations would have biased selection towards nonadditive effects and dominance. They selected the most resistant seedlings (10–20%) in segregating generations. Transgressive segregation was demonstrated in later generations in nine spring-wheat crosses and 17 winter-wheat crosses that lacked resistant progeny in the F_2 and F_3 generations. It is these crosses, without resistance in the early generation, that are of special interest.

Table 8.2 gives an example. The spring-wheat cultivars Manitou and Centana were crossed; selection began in the F_2 generation when there were a few plants with a type 2 reaction; only the F_4 generation did resistant plants (with type 1

*To breed for horizontal resistance Robinson also starts with susceptible parents. Specially noteworthy is his insistence that throughout the breeding program only a single virulence genotype of the pathogen should be used. This makes irrelevant any of the host's vertical resistance genes for which the pathogen is virulent. Using mixtures of pathogen genotypes would bring the danger of selecting combinations of vertical resistance genes which, singly, one or the other of the pathogen genotypes could match. The danger in using mixtures of pathogen genotypes as inoculum in selecting progenies for transgressive resistance is that the selected transgressive resistance could be vertical simply as a consequence of the inoculum being a mixture.

TABLE 8.2

The Percentage of Plants Resistant to *Puccinia striiformis* as a Result of Selecting Resistant Plants and Selfing Segregating Populations of a Cross between the Spring-Wheat Cultivars Manitou and Centana[a]

Generation	Total number of plants	Resistant plants (%)
Parental		0
F_1		0
F_2	96	0
F_3	23	0
F_4	171	8.2
F_5	141	15.6
F_6	177	93.2

[a] Data of Krupinsky and Sharp (1979). Their infection types 00, 0−, 0, 1−, and 1 are taken to indicate resistance, and types 2, 3−, 3, and 4 susceptibility.

reaction or less) appear; but by the F_6 generation most plants were resistant. Results of this sort were consistently obtained.

The Krupinsky–Sharp technique, if we can call it this, is to do without genes of large effect, either by starting with susceptible parents or, alternatively, by discarding all resistant plants in the F_2 and F_3 generations. Adequate resistance can then be expected only in the F_5 or F_6 generation. Krupinsky and Sharp are returning to the idea of polygenic resistance in the true sense, i.e., resistance by genes all of which have only a small effect. This is in clear contrast with the work on, say, resistance to *Helminthosporium turcicum* or *Puccinia sorghi* in maize, because here, on the evidence given in previous sections, much resistance was contributed by a few genes of large effect. It may well be that the wheat lines selected by Krupinsky and Sharp are the first examples of *man-made* polygenic resistance to plant disease. Their experiments will therefore be followed with great interest, especially to see whether events will confirm that the resistance is really stable and horizontal.

The Krupinsky–Sharp technique is particularly apt for crops like wheat that are natural inbreeders and therefore tolerate homozygosity in many loci. For reasons given in Section 8.4, the technique probably demands that all parental material should be agronomically good. This would be no hardship; indeed, it would exploit the common situation in which the popular cultivars are agronomically good but lack adequate horizontal resistance.

The implication in the Krupinsky–Sharp technique, that resistance genes of large effect are unwelcome, is not necessarily and generally true, as maize breeders, among others, well know.

8.11 DIFFERENT METHODS OF ANALYZING VARIANCE

The methods of analyzing variance in Chapters 5 and 8 differ essentially. In Chapter 5 both host and pathogen varied, and there were degrees of freedom for pathogenicity and for host–pathogen interaction. In Chapter 8 these degrees of freedom are absent. In the sort of analysis discussed in this chapter, phenotypic variance is divided into environmental (nonheritable) variance and genetic variance; the genetic variance is divided into additive variance and nonadditive variance; the nonadditive variance is divided into dominance and epistatic variance; and there are ways of estimating the number of genes involved. Nowhere do degrees of freedom for pathogenicity enter. A uniform sample of inoculum is postulated, be it a single isolate, a mixture of races, or natural infection in the field. Consequently there are no degrees of freedom for host–pathogen interaction, and no biometric test for horizontal and vertical resistance.

The analyses in the present chapter, lacking as they do degrees of freedom for host–pathogen interaction, must rely on historical evidence for horizontal resistance. Evidence given in Section 8.8 for the stability of the continuously variable resistance of maize to *Puccinia sorghi* in America, despite the fact that the pathogen has had time and opportunity to vary, is historical evidence that this sort of resistance is horizontal. Its value is not less because it is just historical.

The chapter must be concluded with a caution that information is inadequate on many points. It is often stated or implied that additive resistance is horizontal. Section 8.7 gives some evidence for the statement, which has other attractions. Horizontal resistance in additive (see Chapter 3). Biochemically, additive variance and horizontal resistance (or some forms of it) could have a common background in doses of enzymes. But it would go beyond the evidence now available to say that additive resistance is necessarily horizontal. About nonadditive variance there is also scope for confusion. The stigma of vertical resistance is often implied or stated. Admittedly, the variance of well-known genes of vertical resistance, like the *Sr, Lr, Pg,* or *R* genes, is dominant. But there is no reason to believe that all dominant resistance is vertical. After all, dominance simply means that hybrids are not exactly intermediate between their homozygous parents, and deviations from the exact midpoint do not necessarily imply biochemical upsets.

9

Epidemiology of Resistance to Disease

9.1 INTRODUCTION

Continuously variable resistance is almost inevitably incomplete resistance; and much of what follows is about incomplete resistance.

As an approximation useful for diseases like potato blight and the cereal rusts, which cause explosive epidemics, one can associate vertical resistance with complete resistance to avirulent races during the boom period of the boom-and-bust cycle and horizontal resistance with incomplete resistance to all races. These associations are not definitions; the definitions of vertical resistance and horizontal resistance remain exclusively those given in Chapter 5. And the associations are approximate and subject to exceptions in both directions.

Vertical resistance is commonly complete to avirulent phenotypes of the pathogen. This completeness is more an artifact than a fact. Incomplete vertical resistance would be common (see Section 9.3) if plant breeders allowed it to be so. Instead, they avoid incomplete vertical resistance if they can; in this they are probably correct. Wheat breeders, e.g., are reluctant to release new cultivars unless they give a resistance reaction of type 0 or 1 when challenged by relevant races of rust fungi; the cultivars they produce are vertically resistant at least when they are released, and the resistance is complete because it is selected for its completeness.

The exceptions to the rule that horizontal resistance is incomplete are common and important. Complete horizontal resistance often takes the form of "population immunity" (Vanderplank, 1975, p. 130). This occurs when the progeny/parent ratio does not exceed 1, i.e., when each parent lesion during its lifetime leads on an average to the production of no more than one daughter lesion. The term "population immunity" was introduced against an ecological background, but it is becoming apparent that the phenomenon is important in ordinary, down-to-earth plant breeding for disease resistance. The general mathematical background will be discussed in the next chapter.

9.2 DISEASE PROGRESS CURVES AND RESISTANCE

Working on the approximations that vertical resistance is complete against avirulent phenotypes of the pathogen but absent against virulent phenotypes, and that horizontal resistance is incomplete, Vanderplank (1968) pointed out that vertical resistance would delay the onset of an epidemic and horizontal resistance would be seen to slow the epidemic's progress down. Consider this in more detail, starting with vertical resistance.

Suppose that there are two potato fields, side by side. In the one field the plants have no *R* gene for resistance to *Phytophthora infestans*; they lack vertical resistance to blight. In the other, the plants have the resistance gene *R1;* they have vertical resistance to races (0), (2), (3), . . ., which have no 1 in their designation (see Fig. 5.1). Blight, we suppose, is rare early in the season, because *P. infestans* does not overwinter abundantly; and both fields are healthy at the start. Later, spores arrive from fields that had been infected earlier or from cull dumps of tubers in which *P. infestans* had survived the winter. Of the spores that arrive 99%, we suppose, are of races that cannot infect the *R1* type. They infect only one of the fields: the field without the *R* gene. The other 1% of the spores, we suppose, are virulent for the gene *R1;* to this 1% the two fields are equally susceptible. The field without the *R* gene starts with 100 times as much effective inoculum as the field with the *R* gene. The initial number of lesions per square meter or per hectare of whatever unit one chooses is 100 times as great in the field without the *R* gene as in the field with it. Vertical resistance given by the *R* gene has reduced initial inoculum to 1/100 of what it otherwise would have been. From the initial lesions the fungus starts to increase, and the epidemic in the field with the *R* gene lags behind by the period needed for disease to increase 100 times. The reduction of effective initial inoculum brought about by the *R* gene manifests itself as a delay in the epidemic, and this delay is easily calculated if one knows the average infection rate (for details, see Vanderplank, 1968). The simple calculations assume that the vertical resistance when it occurs is complete, i.e., that phenotypes of *P. infestans* avirulent for the gene *R1* cannot reproduce on potatoes with this gene, which for all practical purposes is an experimental fact.

The delay of an epidemic through vertical resistance is illustrated in Fig. 9.1. Experimental evidence for the delay was given by Vanderplank (1968, pp. 44–45).

As a generalization, horizontal resistance slows an epidemic down. This is illustrated in Fig. 9.2. In a survey of infection by *Phytophthora infestans* in potato fields in the Netherlands three varieties, Bintje, Eigenheimer, and Voran, were compared. These three varieties have no *R* genes; they differ in resistance horizontally (Vanderplank, 1971) with no evidence for vertical effects. The disease progress curves show that Voran had much more horizontal resistance than Bintje. Infection in both these varieties was noticed at the beginning of July, but whereas the foliage of Bintje was destroyed by the beginning of August, some foliage of Voran survived until September.

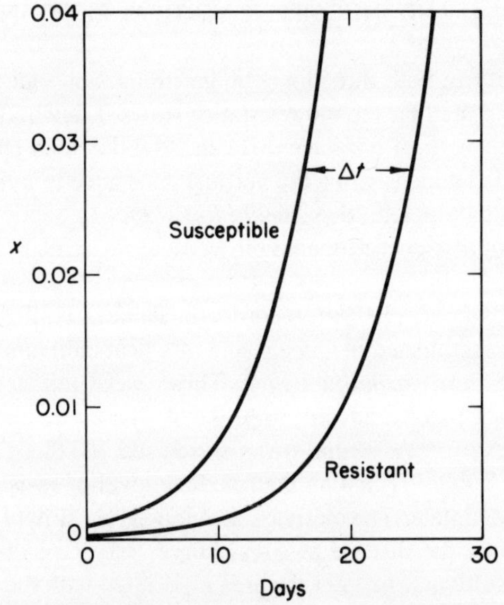

Fig. 9.1. The delay of an epidemic brought about by vertical resistance. The delay Δt is 8 days, i.e., disease at any level is 8 days later in the resistant than in the susceptible variety, when the vertical resistance is effective against 80% of the spores in the initial spore shower, and the infection rate $r = 0.2$ per day. From Vanderplank (1968, p. 43).

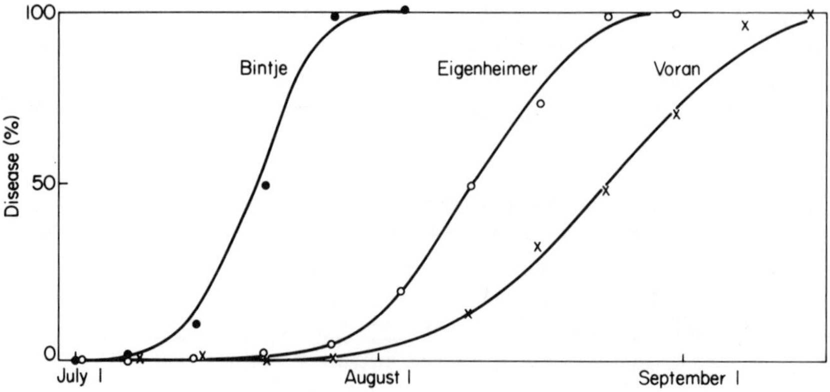

Fig. 9.2. The effect of horizontal resistance. The progress of blight in 117 fields of three potato varieties is shown. The data are for the sand areas of the Netherlands (Anonymous, 1954). The variety Voran has considerable horizontal resistance. From Vanderplank (1968, p. 30).

9.3 SLOW RUSTING AND INCOMPLETE VERTICAL RESISTANCE

One must qualify in both directions the generalization that vertical resistance delays an epidemic, and horizontal resistance slows it down. The first qualification has to do with vertical resistance. It can slow down an epidemic as well as delay its start. This happens when the vertical resistance to avirulent phenotypes of the pathogen is incomplete. See *Partial Resistance* in Section 5.7. In its effect on the disease progress curve incomplete vertical resistance can resemble horizontal resistance.

The data of Rees *et al.* (1979a), recorded in Table 9.1, illustrate this. Experimental plots were established in Queensland of wheat cultivars with intermediate resistance to *Puccinia recondita tritici*. These were infected from artifically inoculated spreader rows, and the progress of disease recorded at successive dates. Two isolates of *P. recondita tritici* were used. To these isolates the wheat cultivars listed in Table 9.1 had an intermediate level of resistance. That is, the resistance was incomplete. The cultivars developed rust slowly. This can be seen from the area below the disease progress curve, relative to more resistant and more susceptible cultivars. In 1971 the area associated with the cultivars in Table 9.1 varied from 24 to 190, compared with 0 for a fully resistant cultivar (Transfer) and 2156 for the exceptionally susceptible cultivar Morocco. In 1972 the figures were 3 to 716, compared with 0 for Transfer and 1278 for Morocco. Ranked according to the area below the disease-progress curve, the cultivars performed differently in the two years (i.e., against different isolates of the pathogen). Much of the resistance was therefore vertical.

TABLE 9.1

Slow Rusting by Two Races of *Puccinia recondita* in Wheat with Vertical Resistance[a]

	Infection type		Adult reaction[b]		Area below curve[c]		Rank[d]	
Cultivar	1971	1972	1971	1972	1971	1972	1971	1972
Warchief	12+		MR-R		24		1	
Hopps	0;2+	X	MS-MR	MR-R	56	3	2	1
Gamenya	0;2−	3+	MR-R	S	63	370	3	6
Pusa 80-5b	2+3++		MS		86		4	
Mengavi	0;2−	3+	MR-R	S	121	716	5	7
Dural	3++	3−c	MR	MR-R	181	36	6	4
Kenya Governor	3+	2	MS	MS-MR	186	121	7	5
Festival	2++	2−n	S-MS	MR-R	190	4	8	2
Warput		3−		MR-R		14		3

[a] From data of Rees *et al.* (1979a). The race in 1971 was race 68-Anz-1,2,3,4, and in 1972 race 76-Anz-2,3. The dates, 1971 and 1972, represent not only different years but also different races.

[b] R, resistant; MR, moderately resistant; MS, moderately susceptible; and S, susceptible.

[c] The area below the disease-progress curve. At the final reading the incidence of leaf rust varied from 30 to 60% in 1972; data were not published for 1971.

[d] The ranking is according to the area below the disease-progress curve.

Another example comes from the work of Clifford and Clothier (1974) on *Puccinia hordei* in barley. The barley cultivar Vada has some resistance to rust, which develops more slowly than in susceptible cultivars. Clifford and Clothier showed that this resistance was at least partly vertical. Using as their criterion the production of spores per cm^2 of inoculated leaf surface, they demonstrated a highly significant differential interaction between barley cultivars and isolates of *P. hordei*, which identifies the resistance as vertical.

Show rusting conditioned by incomplete vertical resistance in cereal crops is probably common. Working with *P. striiformis* Johnson and Taylor (1976) found that there was considerable spore production by pustules classified as type 2− or higher; and that at the higher end of the infection-type scale small differences in classification made for large differences in spore production. Modern varieties of wheat are riddled with genes for vertical resistance to rust fungi. Much of this vertical resistance is incomplete and makes for slow rusting. Slow rusting is not diagnostic of one form of resistance or the other.

It will be remembered from Section 8.10 that the Krupinsky–Sharp technique of selecting transgressive variants was designed specially for the purpose of eliminating vertical resistance even where it is incomplete. Whether the technique will be successful is one of the more important questions in plant breeding for disease resistance.

9.4 HORIZONTAL RESISTANCE THAT DELAYS THE START OF AN EPIDEMIC

The other qualification to the simple generalization has to do with horizontal resistance: It can delay the start of an epidemic, as well as slow the epidemic down after it has started.

There are three ways in which horizontal resistance can delay the start of an epidemic. Two of these have to do with reducing initial inoculum. Commonly, horizontal resistance reduces the percentage of spores that successfully infect. The first cycle of a polycyclic epidemic could therefore be delayed in the same way as vertical resistance delays it. For example, if the percentage of successful infections were halved by horizontal resistance, the start of the epidemic would be delayed by the time needed for disease to double in amount. Only from the second cycle onward would a reduction in the percentage of successful infections reduce the slope of the disease progress curve. The second way in which horizontal resistance reduces initial inoculum is by reducing the general level of disease and hence the level of inoculum that survives from year to year to start a yearly cycle.

The third way in which horizontal resistance can delay the start of an epidemic is by reducing the progeny/parent ratio to 1 or less. This ratio is below the threshold needed for disease to increase. Look again at the disease-progress curve for blight in the potato cultivar Voran in Fig. 9.2. Blight in the foliage of Voran was discovered at the beginning of July, as early as it was in the very susceptible cultivar Bintje. But whereas in Bintje blight developed fast without perceptible delay, in Voran blight stood still for two to three weeks; only after this delay did blight in Voran take off. Voran's resistance to blight is horizontal (Vanderplank, 1971, 1978) and the delay of two to three weeks is an example of how horizontal resistance can delay the start of an epidemic.

The cultivar Voran is typical of many maincrop European potato varieties, about which there is a moderately large literature of resistance against *Phytophthora infestans*. Many workers have studied seasonal changes in blight potential. Very young shoots just emerging from the ground are (in potato varieties without *R* genes) usually very susceptible to artificial inoculation, but there are usually no epidemics of blight at this stage because of the unfavorable microclimate and shortage of inoculum. Later, the plants become more resistant and less easy to inoculate. Finally the plants enter a stage of mature-plant susceptibility, and are easy to infect. The literature goes back right to the 1845 famine in Ireland. Mickle (1845) considered that "... a certain degree of ripeness or maturity is necessary to feed the disease. Early sorts, from all reports, fall the first victims . . . late sorts [in conditions that] tend to retard growth, seem for a time safe, and flatter the possessor they have escaped; but such no sooner reach the apparently required condition, than they may literally be said to wither within an hour." The topic interested de Bary (1876) himself; he considered "... it probable that *Phytophthora* grows more easily on a plant at the height of its development than on young stalks and leaves." Müller (1931) gave a detailed account of changes from extreme susceptibility in the early stages (he considered that the very young shoots of even the most resistant varieties without *R* genes could easily be infected), through an intervening stage of substantial resistance, to a final stage of adult-plant susceptibility. Grainger (1956), who was actively interested in seasonal change in relation to his theory of carbohydrate nutrition, reported that the maincrop variety Kerr's Pink planted at the usual time had a low disease potential for blight in June and July, and a high potential in September. Warren *et al.* (1971) showed that the seasonal change, from susceptibility to relative resistance in midseason back to adult-plant susceptibility was paralleled by changes with age of the leaves. They worked with the potato cultivars King Edward, which is highly susceptible to *P. infestans,* and Arran Victory, which is moderately resistant, neither having an *R* gene. During both the pre- and post-flowering stages, top and bottom leaves were highly susceptible, while the intermediate leaves were relatively resistant and even hypersensitive. Populer (1978) confirmed this. In the moderately resistant potato cultivar Record, which also has no *R* gene, the change of susceptibility in midseason from bottom to top leaves of the stems is sharp. Counted from the bottom, the lower leaves up to leaf 15 on the stem were susceptible; in spray-inoculated plants these leaves developed many lesions, and the lesions were normal. So too above leaf 19 lesions became more frequent and were normal. But on the intermediate leaves, from position 15 through 19, there were few infections, and the lesions were all of the hypersensitive type (see also Section 7.4.) From the work of Warren *et al.* (1971) and Populer we see that the adult-plant susceptibility is probably an adult-field susceptibility, due to the statistical preponderance of adult leaves in the field at that time. Because we are discussing epidemiology, i.e., disease in massed plants in fields, we can accept statistical averages and, with the term adult-plant resistance so widely used in the literature, it is convenient to talk of its counterpart as adult-plant susceptibility.

Our immediate topic is the ability of horizontal resistance not only to reduce the rate of progress of an epidemic, but also to delay its start. The resistance must be adequate. It is adequate for blight in potato varieties like Voran, Record, or Kerr's Pink in Europe, but not Bintje or Eigenheimer, adequacy being judged in relation to climate and microclimate. (In conditions more favorable to blight, as in the Toluca Valley of Mexico, potato varieties like Voran, Record, or Kerr's Pink become blighted without delay.) The resistance in varieties like Voran, without *R* genes, seems to be purely horizontal. This can be judged (Vanderplank, 1971) by the great stability of the resistance over the years. Assessed in the yearly Nederlandse Rassenlijst voor Landbouwgewassen, Voran, licensed in 1936, was 7 in 1938, 7 in 1953, the year of Fig. 9.2 and the heyday of Voran's career, and 7 in 1968, when Voran was fading out as a commercial cultivar. (These are foliage assessments in which 10 = very resistant, and 3 = very susceptible.) By contrast, Bintje was assessed as 3 over those years.

Although the delay in the start of an epidemic of blight in Voran has been discussed here among exceptions to the general rule that horizontal resistance reduces the infection rate whereas vertical resistance delays an epidemic's start, it is, strictly considered, not an exception. It remains an example of the general rule that horizontal resistance reduces the infection rate. Here it reduces the infection rate all the way to zero. To be clear even at the price of being repetitive, this does not mean that the horizontal resistance is enough to prevent successful infection in the field, greenhouse or laboratory. It means that horizontal resistance in an appropriate environment can introduce a population immunity and hence a zero infection rate by reducing the progeny/parent ratio below the threshold of 1 needed for the increase of disease.

9.5 RESISTANCE AS DELAYED ADULT-PLANT SUSCEPTIBILITY

The previous section discussed two phases in the ontogeny of resistance and susceptibility to potato blight. At least in some cultivars there is enough resistance in foliage of middle age to reduce the infection rate to zero or nearly zero; and this is followed by greater susceptibility in adult leaves and hence, on a statistical average, of adult plants and adult fields. In potato cultivars without R genes this middle-age delay of adult-plant susceptibility contributes substantially to the horizontal resistance of the cultivar; indeed, the middle-age resistance is the main component of the horizontal resistance as a whole. Is this a common phenomenon in other diseases?

Rees *et al.* (1979b) set out the examine the behavior of stem rust in slow-rusting wheat varieties. Their data show a pattern very much like that of blight in the potato variety Voran. This is illustrated in Fig. 9.3. They used a single culture of race 21-Anz-2,3,4,5,7 of *Puccinia graminis tritici*. To this race all the varieties included in Fig. 9.3 were susceptible as adult plants, in the sense that they allowed the development of pustules with little or no necrosis or chlorosis. As seedlings, they were also susceptible, with slight variations of reaction recorded in Table 9.2. The plants at the time of inoculation were between the elongation and early boot stages. Assessment of the severity of rust was timed to begin after disease in the most susceptible wheat variety Morocco was clearly visible; and assessments continued at intervals.

Wheat varieties of group G were susceptible over the whole period of the observations. It would therefore be incorrect to talk of adult-plant susceptibility; the susceptibility was unconfined to any particular age. But varieties of groups C and D had almost enough resistance during an intermediate stage of growth to

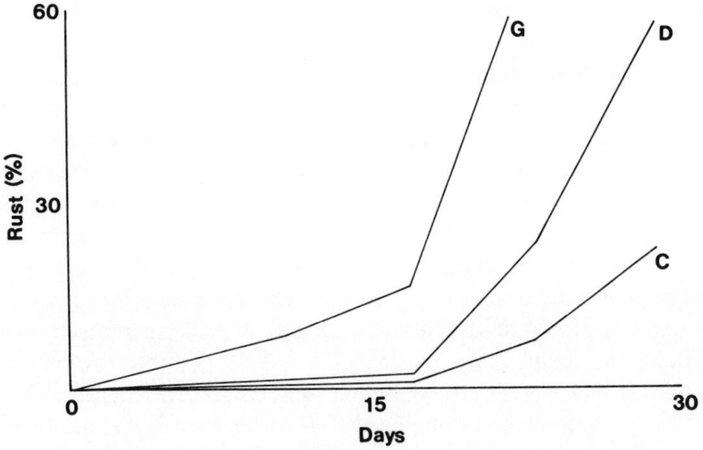

Fig. 9.3. The delay of epidemics of *Puccinia graminis tritici* in wheat cultivars in groups C and D compared with more susceptible cultivars in group G. Zero time in the graph was when rust in the most susceptible cultivar Morocco became clearly visible. The cultivars in the various groups are named in Table 9.2. From the data of Rees *et al.* (1979b).

achieve population immunity during that stage. Only at the end of this stage, in adult plants, did rust increase fast. Here, for wheat stem rust, is evidence for resistance through delayed adult-plant susceptibility; and slow rusting of varieties in groups C and D owes more to this delay than to an unconfined reduction of susceptibility throughout the period of the plant's growth. The pattern of middle-age resistance and adult-plant susceptibility is as clear for stem rust in wheat plants of groups C and D in Fig. 9.3 as for blight in Voran in Fig. 9.2.

Evidence for middle-age resistance and adult-plant susceptibility to rust in wheat, barley, and oats has long been known. Gassner and Kirchhoff (1934) worked with *Puccinia graminis tritici,* and *P. recondita tritici* in wheat, *P. coronata* in oats, and *P. hordei* in barley. They noted that in wheat, oats, and barley varieties with incomplete resistance a wave of resistance just before tillering and flowering is followed by a wave of susceptibility in old plants and the older parts of the plants. This wave of susceptibility in older plants is first seen in leaves in the middle of the stems and later in the top leaves. Gassner and Kirchhoff seem to have been the first to talk of old-age susceptibility (*Altersanfälligkeit*) in cereal rusts. Their evidence for *P. striiformis* was unclear, possibly because of an effect of summer temperatures on this low-temperature fungus. Vohl (1938), working with *P. recondita tritici,* confirmed Gassner and Kirchhoff's findings with this fungus. There was a typical rhythm of field resistance in the common wheat cultivars Marquis, Marquillo, Thatcher, Hope, H44, Garnet, and Reward. Plants of these varieties are susceptible in the seedling stage, begin to become resistant in the shooting stage, are resistant at the time of heading and flowering, and are finally susceptible again in old age. Caldwell *et al.* (1970) reported that the winter-wheat cultivar Vigo, grown on nearly two million acres in 1954, has remained free from severe leaf rust in pure stands in Indiana, despite its being highly susceptible at maturity. It may become severely infected after senescence starts. They also reported that some spring-wheat cultivars have shown promising levels of slow rusting at presenescent stages in the plant's growth even under heavy inoculum. A pattern of infection, wherein early pustules occur mainly on the basal 10 to 20% of the blade, was found in the cultivar Mentana and its derivatives Lerma 50 and 52. With senescence in the host, pustules spread to the distal portion. Such resistance to infection in the cultivar Lerma Rojo 64 and its derivatives gave significant protection to commercial fields in Mexico.

Ohm and Shaner (1976) also observed presenescene resistance to leaf rust, during growth stages 30 to 72, especially 47 to 56, in wheat. Pustule size was markedly reduced, and there was some reduction in the percentage of successful infections per cm^2 after uniform inoculation. They also found an increased latent period in slow rusters, but this is not relevant to a period of population immunity (see Section 10.10). Coupled with the presenescence resistance was an adult-plant susceptibility, i.e., resistance manifested as slow rusting declined after flowering. Ohm and Shaner point out that even though slow-rusting resistance declines with senescence, it may nevertheless provide enough protection to the crop before and during flowering to prevent significant loss of yield from leaf rust. If an epidemic is delayed long enough, the area under the disease progress curve may be insubstantial.

Shaner and Hess (1978) underline adult-plant susceptibility to leaf rust in wheat. They refer to the large increase in rust severity often seen in slow-rusting wheats just before ripening.

9.6 SLOW RUSTING AND HORIZONTAL RESISTANCE

Commonly in the literature slow rusting is equated with horizontal resistance. The equation must be viewed with caution, because some slow rusting is conditioned by resistance that is clearly vertical; this was the topic of Section 9.3. Nevertheless there is enough evidence for optimism that at least some slow rusting, manifested as delayed adult-plant susceptibility, might be conditioned by horizontal resistance.

In part, the evidence comes from analogy. Slow rusting, as delayed adult-plant susceptibility, resembles slow blighting, as delayed adult-plant susceptibility, in the potato variety Voran; and this slow blighting originates undoubtedly in horizontal resistance. In part, the evidence is direct, though not above suspicion. The results of Rees *et al.* (1979b), reproduced in Fig. 9.3 and Table 9.2, would seem

TABLE 9.2

The Infection Type of Various Wheat Cultivars in the Seedling Stage to Infection by Race 21-Anz-2,3,4,5,7 of *Puccinia graminis tritici*[a]

Cultivar	Group[b]	Infection type
Bordan	C	3,3++
Puglu	C	3++
Ford	D	3+
Spica	D	3
WC 306	D	3+
Bunyip	G	3++
Federation	G	3++
Mengavi	G	3+
Morocco	G	4
SHJ	G	3+
Ward's Prolific	G	3++
WC 561	G	4

[a] Data of Rees *et al.* (1979b).
[b] The groups are those shown in Fig. 9.3.

to exclude vertical resistance because only those wheat cultivars were included that gave high infection types, both in seedlings and adult plants. However, one must heed the warning by Johnson and Taylor (1976) that considerable vertical resistance may be hidden within a high infection type. In part, the evidence is historical. It is commonly stated in the literature of the various cereal rust diseases that slow rusting is a varietal character that has remained stable over many years.

9.7 THE INEPTNESS OF SOME INFECTION RATE AVERAGES

The greater the seasonal and ontogenic changes, the less apt as a measure of resistance is the infection rate if it is calculated grossly over a long period.

The infection rate r calculated as

$$r = \frac{1}{t_1 - t_0}\left(\ln\frac{x_1}{1 - x_1} - \ln\frac{x_0}{1 - x_0}\right)$$

correctly gives the average value of the rate at all moments between t_0 and t_1, however much the rate may have varied during that period (Vanderplank, 1963, p. 76). About the arithmetic accuracy of the average infection rate there is no doubt. The doubt is sometimes about the usefulness of the average. For example, curve D in Fig. 9.3 illustrates an infection rate that for 17 days was small, nearly

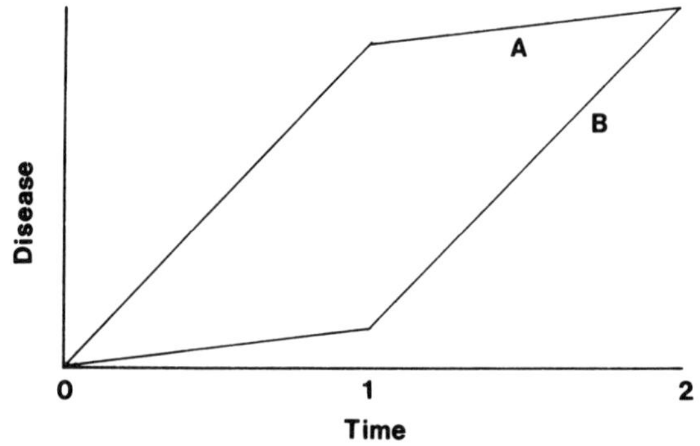

Fig. 9.4. The progress of disease in two very different host varieties, the infection rate averaged between times 0 and 2 being nevertheless the same.

zero, and thereafter was high. Averaged over the whole period of the graph, the rate has little meaning.

Consider Fig. 9.4 illustrating progress curves of disease in two hypothetical varieties A and B. The two disease-progress curves start near zero at time 0 and end at the same level at time 2. Variety A is highly susceptible early in the season, and the infection rate between times 0 and 1 is high; but it has adult-plant resistance, and the infection rate between times 1 and 2 is small. Variety B is the opposite. It has young-plant resistance, and the infection rate between times 0 and 1 is low; but it has adult-plant susceptibility, and the infection rate between times 1 and 2 is high. The disease-progress curves, and the area below them, are very different for the two varieties. Yet the average infection rates, measured over the whole period from time 0 to time 2, are identical, because the two varieties had the same level of disease at time 0 and the same level at time 2. Clearly, an average infection rate estimated grossly over the full period has little use because it does not reflect seasonal and ontogenic change. For the infection rates to the meaningful the varieties should have been compared between times 0 and 1, when B was more resistant than A, and between times 1 and 2, when A was more resistant than B.

Infection rates usefully measure resistance or susceptibility, and the seasonal and ontogenic changes they undergo. Infection rates are essential to modern plant pathology. But averages of highly variable rates must be used with caution and common sense.

9.8 TESTING FOR RESISTANCE AS DELAYED SUSCEPTIBILITY

If resistance is manifested as delayed adult-plant susceptibility, resistance ought to be correlated with the length of the delay, as Rees *et al.* (1979b) found;

and the length of the delay ought to provide a direct method of screening lines for resistance. Small plots of the lines under test could be lightly inoculated from infected spreader rows, those plots that become diseased last being taken as the most resistant.

It would help if the delay in the more resistant lines could be sharply defined by reducing the infection rate during the delay to zero or near zero, i.e., by reducing the progeny/parent ratio to 1 or less. Tests carried out in environments less favorable to disease would be a solution, but are not always practicable. Another solution would be to mix the seed of the lines being tested with seed of distinctive varieties having complete vertical resistance to the inoculum or with seed of other species, such as wheat with barley.

Selection of slow-rusting lines by measuring delayed susceptibility would be less tedious and more integrated than selection by measuring the components of slow rusting: fewer pustules per unit area of inoculated leaf surface, fewer spores produced per uredium, smaller uredia, and a longer latent period. Selection by measuring delayed susceptibility would have the objection that it does not take the latent period into account. This objection may, however, prove to be unimportant, because in slow-rusting barley infected with *Puccinia hordei* Johnson and Wilcoxson (1978) found that a longer latent period, fewer uredospores produced per unit area of leaf surface, fewer spores per uredium, fewer uredia per unit area of leaf surface, and smaller uredia were related to each other and to the area under the disease-progress curve, as indicated by significant correlation coefficients.

9.9 YOUNG-PLANT SUSCEPTIBILITY

Ontogenic opposites to what we have been discussing are young-plant susceptibility and adult-plant resistance. The former is the topic of this section, and the latter of the next.

With the term young-plant susceptibility we associate young-tissue or young-organ susceptibility, as shown for example by the greater susceptibility to *Guignardia citricarpa* of citrus fruits when they are young.

Powdery mildew of barley caused by *Erysiphe graminis hordei* strongly attacks young plants and young leaves. The early literature is cited by Graf-Marin (1934). Judged by the data of Aust *et al.* (1980) analyzed in Table 9.3 infection by a given number of conidia may be 50 times as great in primary leaves in May as in older leaves early in June. Similar results were obtained by Lin and Edwards (1974) and Russell *et al.* (1976), who showed that the change of resistance was expressed at various stages of disease development, including germination of conidia, penetration, formation of elongating secondary hyphae, colony growth, and sporulation.

TABLE 9.3

The Seasonal and Ontogenic Effect on the Average Number of Lesions Produced in Barley Leaves by 1000 Conidia of *Erysiphe graminis hordei*[a]

Date	Leaf[b]	Temperature (°C)[c]	Number of lesions
May 8-10	1	10.8	217
May 16-18	3	11.9	189
May 24-28	5	15.0	143
May 24-28	6	15.0	63
May 30-June 1	6	20.2	12
May 30-June 1	7	20.2	10
June 5-7	7	21.2	4
June 5-7	8	21.2	5
June 12-14	8	12.2	4

[a] Calculated from data of Aust *et al.* (1980), for infection in the field.
[b] 1 = primary leaf; 8 = flag leaf.
[c] Arithmetic mean of the daily temperature.

For pathogen to exploit young-plant susceptibility, it must be able to colonize the susceptible tissue quickly. *Erysiphe graminis* is well adapted to do so. Much inoculum survives the winter, in infected fall-sown crops or as cleistothecia, so that the pathogen is well poised for immediate attack in the spring. Multiplication is fast, because the latent period is unusually short; Jenkyn (1973), working with plants at the 2–3 leaf stage, puts it at 5–8 days at temperatures prevailing in England from the beginning of May until the end of September. Aust and Kranz (1974) found that the time to sporulation after inoculation with the lowest concentration of conidia (which is the most relevant) was 5.3 days at 22°C, 18,000 lx, and 70% relative humidity. These figures are substantially smaller than those for the latent period of the cereal rusts at corresponding temperatures, possibly because colonization by powdery mildew is superficial and therefore quickly established.

A short duration of very high young-plant susceptibility followed by steadily increasing resistance suggests that vulnerability to disease may be strongly influenced by the speed of ontogenic change. The effect of fertilizer on barley powdery mildew, with nitrogen greatly increasing and phosphorus greatly decreasing disease, may be an example of this.

It is just possible that the available evidence for *E. graminis hordei* gives an incomplete picture of the cereal powdery mildews on plants in old age. Old-age susceptibility has been reported in both forms *tritici* and *avenae*. In a study of *E. graminis avenae* in oats, Jones and Hayes (1971) found that each individual leaf exhibited a maximum level of resistance when it was fully expanded, and that thereafter susceptibility increased until the onset of senescence. The general tendency for increased susceptibility in older leaves was evident when a comparison was made of the level of infection on successive leaves and of the amount of mildew on corresponding leaves sampled at successive dates. This was clearly demonstrated when plants sown on April 22, in that leaves 5, 6, and 7 were relatively resistant to infection when inoculated on June 1 but were very susceptible when inoculated on June 23. At this latter date only the upper, most recently developed two leaves (flag and flag −1) showed strong resistance; the older leaves were more susceptible. These observations, of susceptibility in youth, resistance in middle age, and again susceptibility in old age, suggest that there is possibly a similar pattern for three taxonomically very different biotrophs, *Puccinia, Erysiphe,* and *Phytophthora infestans* (which at least initially is biotrophic). Establishing patterns would be important in research, because it is known that resistance to *P. infestans* in potatoes without *R* genes is horizontal. It would also be important for the study of biotrophy.

9.10 ADULT-PLANT RESISTANCE

Adult-plant resistance is defined as resistance absent in young seedlings and developed as the plant matures. It is one of the anomalies of cereal literature that adult-plant or mature-plant resistance is commonly used as a synonym for horizontal resistance, whereas in fact the best-documented evidence is for adult-plant resistance that is vertical.

The wheat varieties Thatcher and Red Bobs are susceptible as seedlings to both races 9 and 161 of *Puccinia recondita tritici* (Bartos *et al.* 1969). As the plants mature, Thatcher becomes resistant to race 9 but not 161, and Red Bobs remains susceptible to both. The genetics of virulence in *P. recondita* on adult plants of Thatcher was studied. In the F_2 progeny of a cross between races 9 and 161 analysis showed that virulence was inherited on a gene-for-gene basis, the adult-plant resistance gene in Thatcher being recessive.

Priestley (1978) has documented the fate of four different adult-plant resistances in wheat to *Puccinia striiformis tritici*. All four have been matched in time by isolates of the pathogen apparently specifically virulent for them.

In maize, resistance to *Puccinia sorghi* often called adult-plant or mature-plant resistance is horizontal, to judge by its history of stability (see Section 8.8), but there is probably a confusion of terms here. Adult-plant resistance, strictly defined, in resistance shown by adult plants but not young plants; but maize pathologists seem to use the term adult-plant resistance to mean field resistance, i.e., resistance assessed in the field and observed as a low amount of disease when the fields are mature. There seems to be nothing in the literature to indicate whether this low amount is the result of resistance in the plants at maturity or in middle age or at any other time or at all times during growth. In maize pathology, adult-plant resistance seems to be a term determined more by technique than by ontogeny. Adult-plant resistance as field resistance, is assessed under conditions of natural infection, and contrasts with seedling resistance, which is resistance assessed in seedlings so heavily inoculated with spores in the greenhouse that only vertical resistance conditioned by *Rp* genes could possibly be manifest.

10

The Anatomy of Epidemics

10.1 INTRODUCTION

A mathematical model quantitatively describes a situation in the real world. To construct it one investigates the real system and tries to develop equations or a computer analysis which mimics the dynamic behavior within the system. The aim of the modeling is a better understanding of the system. The model must not only mimic; it must also incorporate or allow for all the essential variables and parameters within the system.

It is perhaps easiest to illustrate the purpose of models by describing what is not a model. The disease pyramid, much in vogue recently, is not a model. The pyramid shows a time ''dimension'' standing on the base of the disease triangle: host, pathogen, and environment. It is not a mathematical model for two reasons. First, it mimics no biological process known on earth, and therefore contributes nothing to the understanding of plant disease processes. Second, it is mathematically intractable.

Transformations that simplify (linearize) complex data are not models. They include, to give the simplest mathematical example, the square root of the proportion of disease. Fourier analysis has been suggested for disease in perennial plants that comes in yearly waves. All this is mere curve fitting. Most transformations are not models for at least two reasons. They neither explain nor attempt to include obvious components of the increase of disease, such as the latent period. They usually make the fundamental error of assuming that experimental data should be fitted by chosen parameters (i.e., constants). Disease processes are variable; and if one is permitted to use variables in transformation equations there is little difficulty in fitting almost any equation to almost any experimental data. The purpose of these comments is not to disparage the use of transformations, which often are essential, but to draw the line at confusing transformations with models.

10.2 THE LOGISTIC EQUATION IS NOT A MODEL

Many writers still refer to logistic increase of disease. This is incorrect. The logistic equation is used to define logits and develop the useful speedometer known as the infection rate r; but it fails to describe the infection process. It fails among other reasons because it ignores the latent period; and the latent period confounds the logistic equation as a model. If disease were to proceed under absolutely constant and favorable conditions of host, pathogen, and environment, it would not proceed logistically but would proceed increasingly faster than logistically. That is, the infection rate defined by logits would increase with time despite the postulated constancy of conditions, all because of the latent period.

The reasons for rejecting the logistic equation as a model of disease increase were given by Vanderplank (1963, pp. 311–312), and there is no point in going over the same ground again. The logistic equation gave us logits and an infection rate, but no mathematical model to be used in epidemiology.

10.3 THE BACKGROUND TO MODELING

Plant disease is a sequence of processes in time. Fungus spores (or bacterial cells or virus particles) reach the plant. This has conveniently been called the catch. Some of the spores (to keep the illustration to common fungus disease) germinate, and some of the germinated spores complete the various processes needed for them to infect. After infection the fungus establishes itself and develops within the plant to prepare itself for producing spores; the whole period from inoculation to the dispersal of the new crop of spores is the latent period. Spore production, dispersal, and lesion growth continue, sometimes indefinitely, more often until the lesion is "removed," and the period up to removal is the infectious period. In polycyclic ("compound interest") disease the processes are repeated, usually with overlapping; and it is with polycyclic disease that we shall be mostly concerned.

The sequence can be treated analytically or synthetically. For example, in an analytic treatment sporulation can be broken down into a series of events, such as the formation of fruiting bodies, development of sporophores, spore formation, spore maturation, spore liberation, and spore dispersal; infection can be broken down into germination, germ tube development, production of appresoria, and host penetration. Indeed, a specialist would have no difficulty in recommending further subdivision. The various stages can then be related to temperature, leaf wetness, host susceptibility, and all other factors known to be relevant. There are separate models for each particular disease, often relevant only to some particular climate or circumstances. Kranz (1974) listed 14 different equations proposed by various workers to estimate progress for various diseases, and other equations have been proposed since. Many of these equations are purely empirical, and do not cope with the mathematical problem of allowing for a steadily shrinking supply of infectible tissue as the epidemic progresses.

In the synthetic approach use is made of the fact that many processes in the infection sequence have equivalent effects. For example, except at the beginning and end of a polycycle, doubling the catch of spores, doubling the proportion that germinate and infect to start a lesion, or doubling spore production in a lesion would have equivalent effects.

Basically there are two synthetic models of epidemics. In the one, the rate of increase of disease is proportional to the amount of infectious tissue and to the amount of tissue still uninfected but available for infection. The whole is dealt with in a single differential equation which involves both the latent period and the infectious period. This is the model discussed in detail in this chapter. The other model splits the differential equations, and replaces the infectious period by a rate of removal of infectious tissue. Both models have their pros and cons, which are discussed briefly in Section 10.14.

The three essentials in the synthetic model are the initial inoculum, the progeny/parent ratio, and the latent period. The initial inoculum, in relation to sanitation, vertical resistance, and the delay in the start of an epidemic, has been widely discussed in recent years, and details of the discussions need not be repeated here. For the purpose of this chapter, the initial inoculum is the constant of integration for the equation used in calculating various tables; and in order to concentrate attention on the progeny/parent ratio and the latent period, the constant of integration has been kept very small, near zero. The latent period, which is the time newly inoculated tissue needs to become infectious, has also been widely discussed in recent literature. It plays a considerable part in disease especially when the progeny/parent ratio is large. The progeny/parent ratio is the average number of progeny per parent infection. It is the average number of daughter lesions per parent lesion, or for systemic disease it is the average number of plants infected by a single diseased plant during its lifetime. These averages are for the logarithmic stage of the epidemic, when infection is unrestricted by any shortage of healthy tissue available for infection; measurements at later stages must be transformed appropriately.

10.4 THE PROGENY/PARENT RATIO IN REALITY

In later sections the progeny/parent ratio will be considered as an entry in equations, and it would be as well now to demonstrate it as a reality.

Figure 10.1 uses data of Underwood *et al.* (1959) for the "simple interest" phase of an epidemic in wheat caused by *Puccinia graminis tritici*. Wheat plants of the variety Knox were artificially inoculated when they were 20 to 25 cm tall and in the tillering stage. "Flecking" was first noticed 9 days after inoculation, and 1 or 2 days later, on May 3 and 4, the pustules ruptured the host tissues and began to shed uredospores. For the next 13 days the number of pustules remained constant, i.e., the pustules were the original ones developed directly by artificial inoculation. They were estimated at 1.2% of the culm, i.e., $x(0) = 0.012$, where $x(0)$ is the proportion of disease at time 0 in the graph. Thereafter, from May 17, secondary infection became evident with the appearance of new

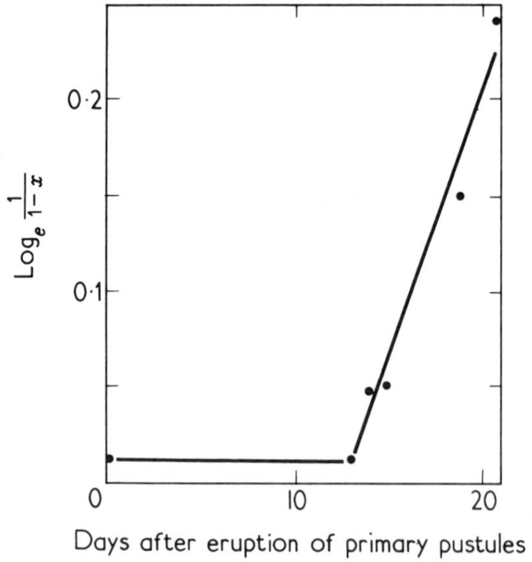

Fig. 10.1. The increase of wheat stem rust caused by *Puccinia graminis tritici* with time. $\log_e [1/(1-x)]$ is plotted against the number of days after the eruption of the primary pustules formed by artificial inoculation. The figure can be used to estimate the progeny/parent ratio. Data of Underwood *et al.* (1959), from Vanderplank (1963, p. 45).

flecks and pustules. Their origin was the crop of uredospores released by the primary pustules from May 8 onward, conditions from May 2 to 7 being unfavorable for infection. Figure 10.1 follows the course of this secondary infection for 8 days, which on internal evidence was slightly less than the latent period. Because the amount of infection at the end was well outside the logarithmic phase, x was transformed to $\log_e[1/(1-x)]$, which was 0.242 at the end. This gives a minimum estimate of the progeny/parent ratio of 0.242/0.012 = 20. This estimate, 20, is a minimum mainly because there was no internal evidence to show how long the infectious period extended beyond 8 days. However, in relation to the tables which follow in the next section a minimum estimate serves its purpose adequately.

A simple example is given by the data of Asai (1960) because the amount of disease was small enough to need no transformation. The data were again for wheat stem rust. After artificial inoculation of wheat by *P. graminis tritici* on June 15 pustules developed and stayed at a constant level of 0.24 pustules per culm for 10 days until July 3. Thereafter they increased, and 6 days later, on July 9, there were 13.65 pustules per culm. The progeny/parent ratio was thus at least 13.65/0.24 = 57, the true figure possibly being substantially higher.

10.5 THE PROGENY/PARENT RATIO IN AN EQUATION

The essence of our model is to equate the rate of increase of disease to the amount of diseased tissue that is infectious and to the amount of healthy tissue remaining to be infected. If x is the proportion of diseased tissue (or the proportion of plants systemically infected), p is the latent period, and i the infectious period.

$$dx(t)/dt = R[x(t-p) - x(t-i-p)][1 - x(t)] \tag{10.1}$$

Here $x(t)$ is x at time t; it is the proportion of disease at all stages of development at time t. Similarly $x(t-p)$ and $x(t-i-p)$ are x at times $t-p$ and $t-i-p$, respectively; $x(t-p)$ is the proportion of tissue that has been infected for at least as long as the latent period and therefore is or was infectious; $x(t-i-p)$ is the proportion that has been infected for at least the latent and the infectious period and therefore was infectious but has ceased to be so; $x(t-p) - x(t-i-p)$ is therefore the proportion of infectious as distinct from infected tissue; $1 - x(t)$ is the proportion of healthy tissue still available for infection; and R is the basic infection rate. The equation is a useful definition of the infection rate which finds many applications.

To put the equation in terms of the progeny/parent ratio α we adopt i as the unit of time, where

$$dx(t)/dt = \alpha[x(t - p) - x(t - I - p)][1 - x(t)] \quad (10.2)$$

Here $\alpha = iR$, and is the progeny/parent ratio, and t and p are in units of i.

Disease starting from very low levels, i.e., with $x(0)$ very small, and increasing according to Eq. (10.2) will eventually level off at an asymptote L where

$$L \sim 1 - \exp(-\alpha L) \quad (10.3)$$

This approximation becomes increasingly accurate the smaller the constant of integration $C = x(0)$ is. Some values of L calculated by Eq. (10.3) for different values of α are given in Table 10.1.

Table 10.2 records values from a numerical integration of Eq. (10.2) when $\alpha = 2$ and $p = i$. Note that disease reaches an upper limit (asymptote) of 0.797, which is also the value calculated for $\alpha = 2$ in Eq. (10.3) and entered in Table 10.1. So too Tables 10.3 and 10.4 record values when $\alpha = 1.5$ and 1.2, respectively, and the upper limits reached 0.582 and 0.314; and these again tally with the corresponding entries in Table 10.1.

TABLE 10.1

The Progeny/Parent Ratio in Relation to the Upper Limit of Disease Which Starts from Very Low Levels[a] and Has Unlimited Time to Develop

Progeny per parent lesion[b]	Disease at upper limit[c]	
1	0	
1.01	0.019	
1.1	0.176	
1.2	0.314	See Table 10.4
1.5	0.582	See Table 10.3
2.0	0.797	See Tables 10.2 and 10.5
3	0.940	
4	0.980	
5	0.993[d]	
20	>0.999	See Table 10.6
∞	1.00	

[a] Disease near the limit of zero. For this reason the first entry in the second column is 0.

[b] This is α of Eq. (10.3); it is the average number of lesions produced during its whole lifetime by one parent lesion, when there is unlimited healthy tissue available for infection.

[c] This is L of Eq. (10.3), disease being expressed as a proportion; e.g., $L = 0.019$ means that there is 1.9% disease at the upper limit.

[d] An illustration of the integration of Eq. (10.2) with $\alpha = 5$ and $L = 0.993$ was given by Vanderplank (1965, Fig. 3, curve C).

TABLE 10.2

The Progress of an Epidemic According to Eq. (10.2) When $\alpha=2$, and $p=i=1$

Time[a]	A Infected tissue[b]	B Infectious tissue[c]	100B/A
12	0.0255	0.0060	23
14	0.067	0.0165	24
16	0.163	0.039	24
18	0.341	0.079	23
20	0.558	0.111	20
21	0.644	0.106	16
22	0.706	0.086	12
23	0.745	0.062	8
24	0.769	0.039	5
25	0.782	0.024	3
26	0.789	0.013	2
27	0.793	0.007	0.9
28	0.795	0.004	0.5
29	0.796	0.002	0.25
30	0.797	0.001	0.13
31	0.797	0.001	0.13
32	0.797	0.000	0

[a] Time is in units of i, and entries until $t = 12$ are not shown.
[b] This is $x(t)$, where the constant of integration $C = x(0) < 10^{-5}$.
[c] This is $x(t-p)-x(t-1-p)$ of Eq. (10.2).

10.6 THE EFFECT OF DWINDLING INOCULUM

Consider Table 10.2 in more detail. Disease (column A) progressed relatively fast with time until time 20 was reached; disease was then 55.8%, and infectious (sporulating) tissue was at its highest, 11.1% of the total susceptible tissue or 20% of the total infected tissue. Thereafter the amount of infectious tissue started to fall and eventually dwindled to nothing when 79.7% of the tissue was infected.

When disease stopped increasing at 79.7% there was still 20.3% of healthy susceptible tissue available for infection. Disease stopped increasing, not because there was no more healthy tissue available to infect, but because there was no inoculum left to infect it. The epidemic ran out of spores.

So too in Table 10.3 for a progeny/parent ratio of 1.5, disease stopped increasing when 58.2% of the susceptible tissue was infected, the remaining 41.8% remaining healthy despite being susceptible, simply because there were no more spores to infect it. In Table 10.4, for a progeny/parent ratio of 1.2, the corre-

TABLE 10.3

The Progress of an Epidemic According to Eq. (10.2) When $\alpha=1.5$, and $p=i=1$

Time[a]	A Infected tissue	B Infectious tissue	100 B/A
20	0.0172	0.00313	18
22	0.030	0.0054	18
24	0.049	0.008	17
26	0.081	0.014	17
28	0.127	0.021	17
30	0.191	0.031	16
32	0.271	0.039	14
34	0.354	0.042	12
36	0.430	0.042	10
38	0.488	0.031	7
40	0.527	0.022	4
41	0.541	0.017	3
42	0.552	0.014	3
43	0.559	0.011	2
44	0.565	0.007	1.2
45	0.570	0.006	1.1
46	0.573	0.005	0.9
47	0.576	0.003	0.5
48	0.578	0.003	0.5
49	0.580	0.002	0.3
50	0.581	0.002	0.3
51	0.581	0.001	0.2
52	0.582	0.000	0
54	0.582	0.000	0

[a] The first entry is for $t = 20$; for all other details the footnotes to Table 10.2 apply.

sponding figures are 31.4% diseased and 68.6% escaping disease through lack of inoculum.

Table 10.1 shows that only when the progeny/parent ratio exceeds about 5 is a pathogen able to use practically all the susceptible tissue at its disposal.

Maximum levels of disease have been well documented. Analytis (1973) found no incidence of apple scab, caused by *Venturia inaequalis,* exceeding 36% disease of the leaf. Kranz (1975, 1977) studied several diseases, many of which reached only low levels before progress was reversed and the diseases started to decline. His results will be taken up again in Section 10.9. With insight, he attributed the failure of disease to increase beyond the maximum to dwindling inoculum; the term "dwindling inoculum" used in the heading of this section is his.

The equations and tables we have discussed give an oversimplified picture of

TABLE 10.4

The Progress of an Epidemic According to Eq. (10.2) When $\alpha = 1.2$, and $p=i=1$

Time[a]	A Infected tissue	B Infectious tissue	100B/A
42	0.0128	0.00130	10
50	0.032	0.0033	10
55	0.058	0.006	10
60	0.095	0.009	10
65	0.142	0.009	6
70	0.189	0.010	5
75	0.232	0.008	3
80	0.264	0.006	2
85	0.286	0.004	1
90	0.300	0.002	0.7
95	0.308	0.002	0.7
98	0.312	0.001	0.3
99	0.313	0.001	0.3
100	0.314	0.001	0.3
102	0.314	0.000	0

[a] The first entry is for $t = 42$; for all other details the footnotes to Table 10.2 apply.

disease progress to a maximum, in that for heuristic purposes they involve constant values of the progeny/parent ratio. In reality, maxima are probably more often reached in a system of falling susceptibility of the host plant, as was discussed for barley powdery mildew in the previous chapter, or in environments becoming less favorable to disease. But the principle remains the same: a maximum is reached because the progeny/parent ratio is or becomes too low and inoculum dwindles to zero.

The infection rate, as such, does not determine the upper limit of disease. For example, *Erysiphe graminis* in barley often has a very high infection rate but a low upper limit to disease, whereas potato virus X has a slow infection rate but, unless artificially checked in a program to control it, proceeds to 100% infection. (It is a matter of history that many old potato cultivars like Green Mountain, Irish Cobbler, and Russet Burbank were found to be 100% infected when the virus was first discovered.) *Erysiphe graminis* in barley often has a very short latent period (see Section 9.9), which contributes to a high infection rate but is irrelevant to the upper limit of disease; it has a low progeny/parent ratio when with increasing maturity barley becomes increasingly resistant to infection (see Table 9.3), and when infection itself shortens the infectious period by causing leaf blades to absciss. Potato virus X has an unlimited infectious period through clonal propagation unless it is artificially checked, and an unlimited infectious period means an infinitely high progeny/parent ratio which, despite a slow infection rate, allows disease in time to reach 100%.

10.7 THE ROLE OF THE LATENT PERIOD

The latent period affects the *rate* at which the upper limit of disease is approached, but does not affect the *amount* of disease at the limit. The amount of disease at the limit is determined by the progeny/parent ratio and the initial inoculum. The role of the latent period is restricted to determining the capitally important time scale of the epidemic. This can be seen without recourse to mathematics. Of the components that determine the structure of an epidemic,

TABLE 10.5

The Progress of an Epidemic According to Eq. (10.2) When $\alpha=2$, and $p=2i=2$

Time[a]	A Infected tissue	B Infectious tissue	100B/A
20	0.0255	0.00355	14
24	0.076	0.0108	14
28	0.199	0.027	14
32	0.421	0.053	13
33	0.482	0.059	12
34	0.542	0.062	11
35.0	0.5947	0.061	10
35.05	0.5972		
36	0.641	0.060	9
37	0.680	0.053	8
38	0.710	0.046	6
39	0.733	0.039	5
40	0.751	0.030	4
41	0.764	0.023	3
42	0.774	0.018	2
43	0.781	0.013	2
44	0.785	0.010	1
45	0.789	0.007	0.9
46	0.792	0.004	0.5
47	0.793	0.004	0.5
48	0.794	0.003	0.4
49	0.795	0.001	0.1
50	0.796	0.001	0.1
51	0.796	0.001	0.1
52	0.797	0.001	0.1
60	0.797	0.000	0

[a] The first entry is for $t = 20$; for all other details the footnotes to Table 10.2 apply.

namely, the initial inoculum, the progeny/parent ratio, and the latent period, only the latent period is measured in units of time, and it has no units other than those of time. What our model predicts, common sense confirms. For a better understanding, examine Tables 10.1, 10.2, and 10.5.

Table 10.1 makes no mention of a latent period, i.e., the latent period is irrelevant to the contents of the table, which concerns the upper limits of disease alone. Tables 10.2 and 10.5 bring out the role of the latent period numerically. The latent period used in the numerical integration recorded in Table 10.5 is twice that used in Table 10.2; $p = 2$ in Table 10.5, and $p = 1$ in Table 10.2. In both tables the same upper limit, 0.797, is reached, which confirms that the latent period does not determine the upper limit. But in Table 10.5 the maximum is approached more slowly, i.e., the infection rate is lower. Thus, in approximate figures, it took 10 units of time, from 28 to 38, for disease to increase from 20% to 70% in Table 10.5, but only 5 units, from 17 to 22, for the same increase in Table 10.2.

10.8 EPIDEMICS WITH HIGH PROGENY/PARENT RATIOS

In Section 10.4 minimum estimates were made for the progeny/parent ratio in wheat stem rust; they were 20 and 57 for the data of Underwood *et al.* (1959) and Asai (1960), respectively. Asai's epidemic was the faster, but there is no evidence to suggest that it was usually fast for a stem rust epidemic. Table 10.6 uses the ratio 20.

Contrasted with previous tables, Table 10.6 differs markedly in two respects. First, disease proceeds to practical completion, with more than 99.9% of the susceptible tissue diseased. Second, the amount of infectious tissue continues to increase even after 99.9% of the tissue is diseased. The contrast is clearly seen in the last column of the tables. In Tables 10.2 to 10.5, the increase of disease is stopped by dwindling inoculum even when there is much susceptible tissue still available for infection; in Table 10.6 with a high progeny/parent ratio, the increase of disease is stopped by a dwindling residue of susceptible tissue left for the pathogen to infect, even when there is much inoculum still about. Tables 10.2 to 10.5, on the one hand, and Table 10.6, on the other, contrast how an epidemic can be stopped by dwindling inoculum and dwindling susceptible tissue, respectively. Tables 10.2, 10.3, 10.4, and 10.6 all use precisely the same equation as a model, and differ only in the progeny/parent ratio. They reflect the importance of the ratio in epidemiology.

The sort of epidemic illustrated by Table 10.6 is well known in fact. Full epidemics of rust in susceptible varieties of wheat end with abundant uredospores and sporulating pustules. Potato fields recently destroyed by blight have festering lesions abundantly sporulating.

TABLE 10.6

The Progress of an Epidemic According to Eq. (10.2) When $\alpha = 20$, and $p=i=1$

Time[a]	A Infected tissue	B Infectious tissue	100B/A
8.0	0.067	0.0073	11
8.5	0.181	0.021	12
9.0	0.430	0.059	14
9.5	0.792	0.157	20
10.0	0.983	0.363	37
10.2	0.997	0.474	48
10.4	0.999	0.573	57
10.6	>0.999	0.635	64
10.7	>0.999	0.641	64
10.8	>0.999	0.630	63
11.0	>0.999	0.553	55
11.2	>0.999	0.423	42
11.4	>0.999	0.277	28
11.6	>0.999	0.146	15
11.8	>0.999	0.058	6

[a] The first entry is for $t = 8$; for all other details the footnotes to Table 10.2 apply.

High progreny/parent ratios are not confined to explosive epidemics. They occur, for example, with systemic diseases which like tobacco mosaic are not lethal. Here the high ratio stems from an indefinitely long infectious period.

Earlier work on epidemiology (Vanderplank, 1963) tacitly assumed high progeny/parent ratios, and was applicable to diseases like the cereal rusts, potato blight, or tobacco mosaic. That is, it was assumed that epidemics would approach 100% disease except when they started too late or were too slow and were cut off when the fields ripened or when adverse weather intervened. For such diseases logits could aptly be used for the transformation of data if disease was polycyclic, or the multiple infection transformation if the disease was monocyclic.

10.9 EPIDEMICS WITH LOW PROGENY/PARENT RATIOS

Diseases with low progeny/parent ratios reach an upper limit below 100%. Examples are given in Table 10.7, based on data by Kranz (1977). Barley powdery mildew, e.g., reached a peak in May, and thereafter declined. As has already been said, Kranz rightly attributed the end of the epidemic to dwindling inoculum. He found that in the decline of disease dilution by new growth of the

TABLE 10.7

Maximum Levels of Disease in Some Host–Pathogen Combinations[a]

Host	Pathogen	Year	Maximum disease (%)	Date of maximum
Rumex	*Ramularia*	1969	7.9	Aug. 8
		1970	11.6	July 29
		1971	5.7	July 27
Fragaria	*Ramularia*	1970	4.4	Sept. 19
Hordeum	*Erysiphe*	1971	4.0	May 28
Plantago	*Oidium*	1970	15.6	Sept. 22
		1971	52.8	Sept. 21
Tussilago	*Puccinia*	1970	34.5	Oct. 6
		1971	1.7	Oct. 5

[a] From data of Kranz (1977). The pathogens were in sequence *Ramularia rubella, R. tulasnei, E. graminis, Oidium* sp., and *P. poarum*.

host plants was of minor importance. Of major importance was the shedding of infected leaves. This shedding, it should be added, reduces the infectious period, especially if the pathogen is biotrophic, and therefore the progeny/parent ratio.

For diseases with low progeny/parent ratios the epidemiology of 1963 is not wholly satisfactory. Logits are no longer apt for transformation. However, conclusions drawn for the logarithmic phase of epidemics remain unaltered.

A feature of disease with a low progeny/parent ratio is its sensitivity to factors that alter the ratio. Even minor changes in the ratio have large effects; and in the control of disease efforts should be concentrated on reducing the ratio.

10.10 THE THRESHOLD CONDITION FOR AN EPIDEMIC

From Eq. (10.3) and Table 10.1 it is seen that no epidemic will start until $\alpha >$ 1. This condition was stated long ago (Vanderplank, 1963, p. 102).

The latent period does not enter into the condition, which is concerned only with the progeny/parent ratio. When the ratio exceeds 1 the latent period does not influence the upper limit that disease will reach, but is not eliminated as a factor in disease; it does at least influence the rate at which this limit is reached. However, when the ratio does not exceed 1, the latent period loses its whole effect. Mathematics apart, the logic is clear. If disease is not increasing, it does not matter how many cycles it goes through.

The statement that an epidemic will not start until the ratio exceeds 1 needs no qualification. However, if an epidemic has been proceeding with a ratio greater than 1 and the ratio then falls to 1 or less, the epidemic will stop but not immediately. The past history of developing disease has its effect. This is illustrated in the next section.

10.11 A VARYING PROGENY/PARENT RATIO

For heuristic purposes, to determine the effect of the progeny/parent ratio, the ratio was kept constant throughout each of Tables 10.1 to 10.6, and the effect was obtained by comparing the tables. However, the ratio is variable, just as much as its constituents like spore germination and spore production are variable, and its variability is just as much subject to experimental determination as that of its constituents. Also, in a model the ratio can be varied as often as is required.

Table 10.8 illustrates a variation. It is assumed that an epidemic proceeded with a progeny/parent ratio of 2, as in Table 10.2, until disease reached a level of 30%. A protectant fungicide was then applied that kept the ratio down to 1. If the fungicide had been applied at the very start, before the initial inoculum arrived, no epidemic would have developed. But the delay in applying the fungicide allowed disease to build up, and despite a ratio of 1 the disease continued to increase from 30.0 to 49.4% before dwindling inoculum brought it to a stop. (To the observer the increase of disease after the fungicide was applied would seem greater, because at 30% the disease would have included much that was still incubating and invisible, whereas in the end all disease would have been visible.)

Viewed in terms of fungicidal action, the table illustrates the established principle that applications of protectants should start before the expected

TABLE 10.8

The Effect of Varying the Progeny/Parent Ratio by Applying a Fungicide during an Epidemic[a]

Time	A Infected tissue	B Infectious tissue	110B/A
17.6	0.300	0.070	23
18	0.321	0.079	25
20	0.416	0.057	14
22	0.461	0.027	6
24	0.481	0.012	2
26	0.489	0.004	0.8
28	0.493	0.002	0.4
30	0.494	0.000	0

[a] Throughout $p = i = 1$; and $\alpha = 2$ until $x(t) = 0.300$, when the fungicide was applied and α reduced to 1. The time scale is aligned with that in Table 10.2, on to which Table 10.8 is grafted.

epidemic. It is easier to control an epidemic by reducing the progeny/parent ratio to 1 while inoculum is still external than after inoculum has become internal. The greater difficulty with late applications arises from the substantial proportion of built-in infectious tissue which has to be mopped up. The more effective the fungicide, the faster is the mopping up.

The table also illustrates the smaller effectiveness of protectant fungicides compared with eradicant (curative) fungicides. Eradicant fungicides go to the core of the problem and destroy the built-in infectivity.

Endless possible permutations to tables like Table 10.8 are needed to fit the variety of effects of fungicides on epidemics. The scope is unlimited, but a wide discussion is out of place here.

10.12 INTERNAL CHECKS OF ACCURACY

The accuracy of Tables 10.2 to 10.6 can be checked externally and internally. The external check is the agreement with Table 10.1; numerical integration ends with the correct integral. Thus, Table 10.2 ends with the amount 0.797, which appears also in Table 10.1 in the appropriate place. The internal check can be carried out in the tables themselves. Consider Table 10.5. Here $p = 2i$, and time is recorded in units of i. Take $t = 35$ for investigation. From the table, $x(t) = 0.5947$. For $t = 33$, two units of time back, $x(t - p) = x(t - 2) = 0.482$. For $t = 32$, three units of time back, $x(t - 1 - p) = x(t - 3) = 0.421$. From these entries, $x(t - p) - x(t - 1 - p) = 0.482 - 0.421 = 0.061$. This is the figure entered in column B for $t = 35$. In all the tables the entries in column B are consistent with those in column A. (In the tables other than Table 10.5, $p = i = 1$, so that the entries in column B for time t are the differences between the entries in column A for times $t - 1$ and $t - 2$.)

Eq. (10.2) changes to the approximation

$$- \Delta \ln [1 - x(t)] \sim \alpha[x(t - p) - x(t - 1 - p)] \Delta t \qquad (10.4)$$

where Δ means a small interval (difference). The smaller the interval the better the approximation. Consider the approximation, using the entries for $t = 35.0$ and $t = 35.05$ in Table 10.5. The left-hand side of the approximation is the difference between $-\ln (1 - 0.5972)$, which is 0.9094, and $-\ln(1 - 0.5947)$, which is 0.9032, the difference being 0.0062. The right-hand side is $2 \times 0.061 \times 0.05 = 0.0061$. (Here $\alpha = 2$;0.061 comes from the previous paragraph; and $\Delta t = 0.05$.) The approximation, with 0.0061 compared with 0.0062, is adequate for a two-digit check. A comparison of this sort can be carried out anywhere in the tables. Although limitations of space have necessarily restricted the entry of details, interpolation by any suitable formula will provide the data needed for testing.

Tables 10.2–10.6 pass the tests, external and internal, well. By contrast, CERCOS fails badly. Berger (1977), in an otherwise stimulating review, discusses the epidemic simulator CERCOS. Resistance to disease is simulated by lengthening the latent period from 10 days in the susceptible variety to 20 days in the resistant variety. Alternatively, resistance was expressed as a reduction by 50% in the spore catch, percentage of successful infections, lesion size, and spore production. According to CERCOS neither of these forms of resistance can reduce the infection rate when once there is more than 1% infection in the field. Let us examine CERCOS internally, and consider the latent period. According to CERCOS for disease levels above 1%, disease was increasing sixfold in 10 days, e.g. from 1 to 6%, irrespective of whether the latent period was 10 or 20 days. (This figure can be obtained from CERCOS either by reading logits or calculation from the infection rate *r*.) This of course is the basis of CERCOS' statement that the latent period does not influence the infection rate after disease exceeds 1%. It follows that the susceptible variety would have six times as much infectious tissue (i.e., infected tissue that was at least 10 days old) as the resistant variety (in which infected tissue would have to be at least 20 days old before it was infectious). With spore catch, percentage of successful infections, lesion size, and spore production being equal (by CERCOS' hypothesis), foliage in the susceptible variety would be getting six times as much effective inoculum as in the resistant variety. CERCOS is telling us that increasing effective inoculum sixfold would have no effect on disease. This is not only against the experimental evidence; it is also inconsistent with CERCOS' own assumptions. Moreover, even without going into calculations, there is no reason for an abrupt change, with *r* in the resistant variety increasing threefold, from 0.06 to 0.18, as (according to CERCOS' graph) disease increases above 0.3%; the environment is by hypothesis constant, spore catch, percentage of successful infections, lesion size, and spore production are by hypothesis constant, and the percentage, 99.7, of healthy foliage still available for infection is practically unchanged. Similar inconsistencies are revealed when the other side of CERCOS is examined, with the latent period taken as constant, and spore catch, percentage of successful infections, lesion size, and spore production taken as varying. CERCOS simply cannot bear internal examination.

Computers are useful. They store information; a use was suggested for this in Chapter 3. They calculate fast; their speed is used in the potato blight forecasting system BLITECAST (Krause *et al.*, 1975). They are adapted to reiterative calculations like those in Tables 10.2 to 10.6. They do not make mistakes. But their operators can and do; the results computers turn out are as accurate or nonsensical as their operators let them be. There is a regrettable tendency in plant disease epidemiology to shelter behind computers, and leave it to be inferred that what the computer says is necessarily accurate and needs no further probe. The evidence is otherwise. The author of CERCOS failed to tell the computer how to simulate an epidemic. Aspirant epidemiologists might well start by examining published accounts of plant disease simulation programs. The first inquiry would be whether enough information is disclosed to allow internal tests to be carried out (using if necessary an appropriate interpolation technique). The second would be whether the tests are passed.

10.13 ANALYSIS VERSUS SYNTHESIS

In practical epidemiology one must determine the effect on epidemics of changes in the environment, the host plant, and the pathogen. Temperature, for example, affects wind speed and turbulence, spore deposition, germination, infection, lesion growth, sporulation, spore release, host resistance when this is temperature-sensitive, and pathogen virulence as in Chapter 2; it also had indirect effects, as on relative humidity and dew. Several of these categories, like germination and infection, can be subdivided further, temperature affecting the subdivisions differentially. The potential categories are endless when one includes all the factors such as rain, sunshine, pollen on the leaves, biotic competition with the pathogen, and the like. Some degree of synthesis is inevitable. The question is, how much? Analysts will retain, according to circumstances, certain categories, of which spore catch, percentage of successful infections, and spore production are among the more popular. The synthesist, on the other hand, will note that spore catch, percentage of successful infections, and spore production are all components of the progeny/parent ratio; a variation in any one of them would be a variation in the ratio. Why, he might ask, should one not use the ratio itself?

Shrum (1978) has lauded the analytic and disparaged the synthetic (holistic) approach. It is as well to remember here that the proof of the pudding is in the eating. All we know about the structure of epidemics comes from synthetic approaches. All the workable disease-forecasting systems, like BLITECAST, are highly synthetic. So far, the analytic approach has contributed nothing; and there is a real danger that the torrent of unfruitful narrative about analysis, unless it abates, may turn the general plant pathologist against epidemiology.

Ultimately the choice between synthesis and analysis should be the one that gives the smaller experimental error. If the aim of the exercise is to simulate an epidemic or forecast its progress, and not just to study germination, infection, sporulation, etc., each for its own sake, the question is, how does one reduce the experimental error? A direct estimate of the progeny/parent ratio has one experimental error. Estimates of components of the ratio all have their own errors which must be summed (through their squares). The difficulty of estimating the progeny/parent ratio experimentally will depend greatly on whether the latent period is longer or shorter than the infectious period. There is no point in trying to minimize the difficulty. But there is equally no point in pretending that the ratio's components like percentage spore catch or percentage germination are easy to quantify over the whole plant; and in the long run the best justification for splitting the ratio into components, or for splitting some chosen components further, would be evidence that the experimental errors would be so greatly reduced that even when they are summed there would be a net gain of accuracy.

Synthesis can go no further than to leave epidemics with three variables: the initial inoculum, the progeny/parent ratio, and the latent period. A task for the 1980s is to determine and integrate them experimentally, splitting them only if it serves to obtain better accuracy.

There are formidable difficulties ahead in passing from theory to practice; but they would seem equal in synthetic and analytic systems, and cannot be evaded by switching from synthesis to analysis or *vice versa*. First, growth of the host plants during the epidemic may have to be allowed for in some instances. This is relatively easy to cope with. Second, disease is not randomly distributed, but tends to occur in foci which rather than the lesion may be the unit of disease in epidemics (Vanderplank, 1963, pp. 82–89). Third, disease itself changes important parameters and independent variables. It causes leaves to dry up or drop off, thereby altering the microclimate, so that microclimatically a heavily diseased field often differs from a lightly diseased field of the same age. So too when disease causes rapid leaf drop, the infectious period is likely to depend on the level of disease. These difficulties must be faced with little hope that analysis or synthesis will make them less.

10.14 TWO MODELS: PLATEAUS AND PEAKS

We have been using a model with an infectious period which ends when the infected tissue is removed from the epidemic, as when lesions dry up or otherwise cease to form spores. The latent period and infectious period are taken up together in single differential equation. In this model disease, given time enough, increases to an upper limit or asymptote, and there it stays. That is, disease rises to a plateau, represented, e.g., by an entry of L in Table 10.1.

An alternative model, which has been used in medical epidemiology, uses a removal rate instead of an infectious period. There are extra differential equations to define the removal rate. In this model, disease, given time enough, increases to an upper limit which is a maximum, and thereafter decreases. That is, disease has a peak.

The choice of models depends, among other things, on whether a plant disease epidemic is best seen as disease rising to a plateau or a peak. It would seem that a plateau better represents disease in annual crop plants, and a peak disease in people and, probably, in cycles of disease in perennial crops.

There is fundamental difference between epidemics in plants and in people. In plants, infected tissue with rare exceptions remains infected (though not necessarily infectious). Leaves may fall as a result of infection, which gives the impression of decreasing disease and therefore of a peak rather than a plateau. But the fallen leaves must be considered a casualty of the epidemic and included in the loss from disease. Casualties included, disease rises to a plateau. In people, death apart, the diseased in an epidemic usually throw off the infection, and as a result of having been diseased acquire an active immunity of varying effectiveness, which removes them from the epidemic. For illustration, one may think of a disease like measles in children who have not had preventative vaccination. An epidemic starts when much of the population is susceptible, rises to a peak, and then abates as more and more children get over the disease and become immune to further immediate infection. Through their immunity they are removed from the epidemic; and unlike removals of plant disease they are healthy again and part of the normal population. The epidemic has had a peak and passed it.

Cyclic disease in perennial plant populations rises and falls, often with a wave-like incidence of disease. If in the long run one does not consider losses of years ago as casualties to be reckoned with currently, one would have to think of disease in peaks. So far, no attempt seems to have been made to explain the cyclic peaks in terms of infection rates and removal rates; and we must make it clear that the model described by Eq. (10.2) is not intended for cyclic disease in perennial crops.

10.15 APPENDIX ABOUT THE TABLES

The set of conditions stated at the top of each of the Tables 10.2–10.6 specifies the shape of the curve for the progress of disease. For each set of stated conditions the shape remains the same, provided that the initial inoculum is small enough. The position of the curve in time, i.e., when any given level of disease would be found, depends on the initial inoculum; reducing the initial inoculum would delay the onset of disease, but would not change the shape of the disease progress curve.

Consider Table 10.2 for numerical illustration. From the conditions stated at the top of the table, namely, that $\alpha = 2$, and $p = i = 1$, it can be calculated that halving the initial inoculum would delay the onset of disease at any given level by 1.48 units of time. Every entry in columns A and B (and the last column as well) would have been delayed by 1.45 units of time. Thus, in column A the level .558 would have been reached at time 21.48 instead of 20; the level 0.644 at time 22.48 instead of 21; the level 0.706 at time 23.48 instead of 22. So too in column B the level 0.111 would have been reached at time 21.48 instead of 20, and so on. The effect of changing the amount of initial inoculum (provided that the amount remains small) would simply be to raise or lower all the entries in the first (time) column relative to entries in the other columns, without in any way altering the relation between the entries in columns A and B and the last or the time interval between the entries in them.

By keeping the initial inoculum small, we eliminate it as a factor in the shape (as distinct from the position in time) of the disease progress curve. In a footnote to Table 10.2 the initial inoculum, which is the constant of integration, is not specified other than to limit it to amounts less than 10^{-5}. Within this limit it can change without changing the relation between the entries in columns A and B and without changing the time interval between the entries. This limit 10^{-5} is related to the accuracy of the estimates in the table. It is adequate for entries of three digits in column A. If greater accuracy had been desired, with entries in column A calculated to, say, six digits, the limit would have had to be reduced to, say, 10^{-8}. With less accuracy, a higher limit would have been permissible. It is this that concerns us. It is far too optimistic to expect that our model would mimic a real epidemic correct to three digits at each entry. It would be too optimistic even to expect the model to mimic a real epidemic correct to two digits at each entry. With the error of mimicry high, exact values need not be assigned to the initial inoculum. This is why variation in initial inoculum has been ignored in this chapter. Until such time as we know the error of mimicry, it is better to concentrate attention on the progeny/parent ratio and the latent period.

With the entries in the time column able to be moved up or down without affecting the relations between the other columns, Tables 10.2–10.6 give clearer illustrations of the structure of disease progress than anything else yet published, and they show what a wide variety of structures can be explained by the interactions of two variables. If the tables draw attention to the need for further experimental and mathematical research into the progeny/parent ratio, they will have been doubly useful.

11

The Spread of Disease

11.1 INTRODUCTION

This chapter uses the same model, expressed as a differential equation, as the previous chapter, but the background is so different that it is convenient to keep the chapters separate.

Chapter 11 is concerned with the spread of disease as distinct from the dispersal of pathogens. For convenience, we shall use the word *spread* for the movement of disease, and *dispersal* for the movement of pathogens. Such a convention is admittedly arbitrary, but if it removes some of the confusion implicit in the literature, it is well justified.

The spread of disease is much more related to the topic of this book than the dispersal of pathogens. The spread of disease necessarily involves the host plant intimately, and belongs to the study of host/pathogen interactions within the wider scope of host/pathogen/environment interactions. On the other hand, much of the work on the dispersal of pathogens concerns pathogen/environment studies only; indeed, studies of the dispersal of pathogens have drawn heavily on the dispersal of pollen and *Lycopodium* spores without host plants being involved. One need only think of all the work done on the diffusion of a spore cloud in the atmosphere to realize how the emphasis has been on pathogen/environment or organism/environment relations, with the host plant considered as little more than a takeoff or landing strip.

The particular problem of this chapter is the spread of disease outwards from a diseased area, as distinct from the increase of disease within the area, which was the topic of Chapter 10. Linked with this is a study of disease gradients. Facts of the spread of disease have often been recorded in the literature. For example, after blister rust of white pines caused by *Cronartium ribicola* had been well established in the West of North America, the disease front advanced at a rate of about 20 miles a year (Peterson and Jewell, 1968). We shall be more concerned to probe the principles behind the facts.

11.2 BACKGROUND

Few topics in plant pathology have been studied so extensively as the dispersal of pathogens. Nearly 60 years ago Stakman and his associates studied the movement of spores of *Puccinia graminis* in the United States, and showed that early in June 1925 half a million square kilometers of wheat fields were inoculated almost simultaneously (Stakman and Harrar, 1957). Gregory's (1945) review of the literature about the dispersal of airborne spores triggered in the 1950s and 1960s some of the finest work on spore biology. Among topics studied were spore liberation in relation to takeoff in various circumstances, deposition gradients, long-distance dispersal, the dispersal of the spore cloud and the effect of atmospheric turbulence, sedimentation of spores in still air, natural deposition by sedimentation, boundary-layer exchange, impaction, turbulent and electrostatic deposition, and scrubbing by rain. Aylor (1978) has reviewed the literature. Next to dispersal of airborne spores, dispersal of viruses by vectors has probably been the most studied. Thresh (1978) has reviewed this literature, and Harrison (1977) has reviewed the special ecology and control of viruses carried by soil-inhibiting nematodes and fungi. The ring of comprehensive reviews has been closed by Wallace (1978) who reviewed the literature of the spread of soil pathogens.

11.3 THE SPREAD OF MONOCYCLIC DISEASE

The rate of spread of monocyclic disease closely follows the basic infection rate R defined by Eq. (10.1). This in turn is directly related to the proportion of spores that infect, the rate of sporulation, and the spore catch (Vanderplank, 1975, pp. 91–92). Put generally, R is the rate of increase of disease per unit of infectious tissue per unit of time. For monocyclic disease, the increase of disease within an area and its spread to extend the area are closely related. The gradient of dispersal of spores and the gradient of spread of disease are closely related.

Suppose that it takes X acts of dispersal for the pathogen to advance one kilometer. The rate of spread, which is one kilometer in the time needed for X acts of dispersal, is proportional to R. If R is doubled (e.g., by the greater susceptibility of the host plant, or by weather more favorable for infection and sporulation), the frequency of acts of dispersal is doubled, and it will take half as long for X acts of dispersal to occur. That is, if R is doubled, it will take disease half as long to spread 1 km. The rate of spread, other things being equal, will be proportional to R.

$$\text{Rate of spread} \propto R$$
$$= \gamma R$$

Here γ is the proportionality factor. It is γ that (among other things) takes up the dispersal of pathogens. The contrast in roles between γ and R must be understood. Thus, the effect of atmospheric turbulence on a spore cloud is taken up directly by γ and only indirectly by R. The horizontal resistance of the host plants and the aggressiveness of the pathogen, on the other hand, are taken up almost exclusively by R, though there can be an indirect effect on γ if the opening up of foliage by disease affects the gradient of dispersal of spores.

11.4 THE SPREAD OF POLYCYCLIC DISEASE

In polycyclic disease the rate at which disease extends its area is highly geared to the infection rate, and a change of the infection rate r leads to a more than proportionate change in the rate of advance of disease in space. Other things being equal, the infection rate itself flattens or steepens the gradient of spread of disease.

The basic fact of this section is common to all disease, be the pathogen a fungus, bacterium, or virus, or be it dispersed by wind, water, or vectors. Disease at the boundary between disease and no disease is always monocyclic. This is illustrated in Fig. 11.1, which shows an epidemic spreading outward. Inside is an area of polycyclic disease; around this is monocyclic disease; and all around are healthy plants into which disease is in the process of spreading. Countless variations of detail are possible; e.g., disease may advance on an almost straight front. But the area of polycyclic disease is always separated from the area of healthy plants by an area of monocyclic disease. Our discussion is based on this.

Inside the area of polycyclic disease, disease increases at the infection rate r. The outward spread of disease into the healthy area is at a rate proportional to R, for reasons given in the previous section. It does not matter here whether the spores come from the old polycyclic area or from an area previously monocyclic

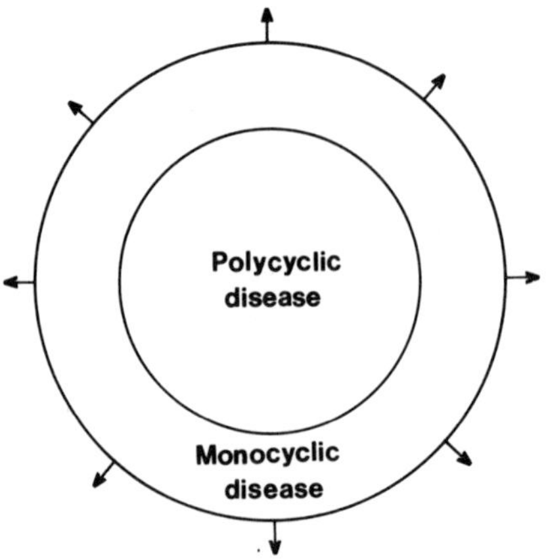

Fig. 11.1. The spread of disease, with monocyclic disease interposed between polycyclic disease and the healthy surroundings.

but now newly polycyclic. The comparison between the rate of increase of disease within the diseased area, i.e., the infection rate, and the rate of spread to increase the area is therefore a comparison between r and R.

Table 11.1 shows the effect of the infection rate on gradients of disease. For the purpose of calculation the latent period p and infectious period i have been kept constant at 10 days. In the first column r increases from 0.0001 through intermediate values to 0.5 per day. The second column records R calculated from r, p, and i by the appropriate equation [Vanderplank, 1975, Eq. (4.3)]. The equation is for the logarithmic phase of increase of disease; and because R is here relevant to disease entering a previously healthy area the use of an equation for

TABLE 11.1

The Effect of the Infection Rate on Gradients of Disease, with $p = i = 10$ Days

r per day	R per day	L	$R:r$	Gradient
0.0001	0.1002	<0.01	1002	Nearly flat
0.001	0.1015	0.03	101.5	
0.01	0.1277	0.40	12.77	
0.1	0.4300	0.99	4.30	Fairly steep
0.2	1.709	>0.99	8.55	
0.3	6.342	>0.99	21.14	
0.4	22.25	>0.99	55.61	
0.5	74.70	>0.99	149.4	Nearly flat

logarithmic increase is permissible. The third column records L, which is the upper limit of disease expressed as a proportion (with $100L$ the percentage of disease). It is defined and estimated by Eq. (10.3) in the previous chapter. For this estimation we need α, the progeny/parent ratio. This is iR, which in this particular table is $10R$ and easily estimated; it is not entered in the table. The fourth column shows the ratio $R{:}r$. To repeat, the rate of spread of disease outward into the healthy area of Fig. 11.1 is proportional to R and the rate of increase of disease within the polycyclic area is r; a high $R{:}r$ ratio therefore means a nearly flat gradient of disease, and a low $R{:}r$ ratio a steep gradient. The last column of the table brings this out.

A general rule is seen in Table 11.1. If the upper limit of disease L is less than 0.99, i.e., if disease will not proceed beyond 99%, the gradient becomes steeper with an increased infection rate. If the upper limit is beyond $L = 0.99$, the gradient becomes flatter with an increased infection rate. In more concrete terms, if the disease is of the sort featured in Table 10.7 of the previous chapter, the gradient will become steeper as the infection rate increases; if the disease is a cereal rust, potato blight, or other disease likely to occur in explosive epidemics, the gradient will become flatter as the infection rate increases. This rule was derived from Table 11.1 with its particular specification of p and i, but other values of p and i can be used without changing the substance of the rule, although there are minor changes of detail.

The point that with very low infection rates the gradient of disease becomes nearly flat can be illustrated by a simile from the countryside. Take the entry at the top of Table 11.1 as an example. With r as low as 0.0001 per day, disease will not increase above $L < 0.01$, i.e., it will remain below 1%. This can be likened to a very low hill in a plain, and because the hill is very low the hillside slopes are necessarily very gradual, i.e., the gradient is nearly flat. With higher infection rates the simile calls for higher hills and therefore steeper hillsides; but beyond a certain point the simile breaks down because the country surrounding the hill is also rising, until finally there is an eminence only slightly higher than its high surroundings.

11.5 THE RATE OF SPREAD OF FAST EPIDEMICS

This section deals only with disease represented in the bottom half of Table 11.1, i.e., with diseases like wheat stem rust or potato blight. As the infection rate increases, disease spreads disproportionately faster, with flatter gradients. Conspicuous foci of disease characteristics of slow infection become less conspicuous as the infection rate increases. It is common experience, e.g., that when the weather is not very favorable for potato blight, and the infection rate is low, pockets of infection are often conspicuous; but when the weather is very favorable, the disease seems to break out explosively here, there, and everywhere almost simultaneously.

A clear distinction must be drawn between the effect of the level of disease and the effect of the rate at which this level is reached. It has long been known that gradients of disease are likely to be steeper when there is 5% disease than when there is 50%; secondary disperal of spores flattens the primary gradients. The point of this chapter is that gradients at the 5% disease level are likely to be flatter when 5% disease is reached at a fast rate than when it is reached at a slow rate. This holds as well for any other level of disease.

This is relevant to the use of fungicides. Applications of protectant fungicides are best begun while most plants are still healthy. Curative fungicides are wasted unless disease is already there. It is relevant to the use both of protectant and curative fungicides to know that disease will be widespread, affecting most plants, at a lower overall average level when the infection rate is high. (This effect is independent of, but supplemented by, the incubation period; a greater proportion of disease is incubating and invisible when the infection rate is high.)

11.6 MONOCYCLIC AND POLYCYCLIC DISEASE

To summarize, monocyclic and polycyclic disease have this in common: They spread at a rate proportional to R. Disease will spread faster if the horizontal resistance of the host plants is small, the aggressiveness of the pathogen is great, and weather is favorable for infection.

Monocyclic and polycyclic diseases differ in this: The gradient of spread of monocyclic disease is likely to follow fairly closely the gradient of dispersal of the pathogen. The gradient of spread of polycyclic disease soon divorces itself from the gradient of dispersal of the pathogen, and takes on a form determined largely by the infection rate.

11.7 THE EFFECT OF THE SCATTER OF DISEASE ON THE INFECTION RATE

Previous sections have dealt with the effect of the infection rate on the spread of disease. Here we reverse cause and effect, and consider how the scatter of disease could affect the infection rate. Direct experimental evidence that is not highly artificial is difficult to get. There is however some evidence which suggests that the scattering of disease increases the infection rate. This could have been expected, because pathogens have mechanisms for relatively long-distance scattering as well as for close dispersal, i.e., in ecological terminology, for "colonizing" as well as for "exploiting." One accepts that Nature knows best.

Croxall and Smith (1976) found that in potato fields the infection rate by *Phytophthora infestans* depended on the amount of infection of the seed tubers that were planted. They analyzed the records of 58 epidemics of potato blight over a period of more than 50 years in the East Midlands of England. They determined from these records the rate at which blight defoliated the crop, by determining the time foliage took to die after the first record of infection. A high infection rate resulted in the foliage being killed within 7 days after the first record of infection; a rate classed as medium resulted in the foliage being killed within 7–13 days after the first record; and a low rate resulted in some foliage persisting undestroyed after 13 days. They also assessed from the records the amount of blight in the stocks from which the seed tubers were drawn. They recorded this assessment of initial inoculum in three categories, high, medium, and low.

Table 11.2 sets out their records in a test of independence in a 3 × 3 classification. There is a highly significant correlation between high initial inoculum and a high infection rate, and between low initial inoculum and a low infection rate. This is especially clear in the table if one compares the extremes of high and low initial inoculum and of a high and low infection rate. Superficially, there is no reason why this correlation should exist. The infection rate reflects the weather during the current season. The initial inoculum reflects the weather during the previous seasons and the conditions during storage the previous winter.

The likely explanation is that widely scattered inoculum promotes fast infection. Consider the diseased tubers used as seed; they contributed in large measure

TABLE 11.2

An Analysis Showing the Relation between the Amount of Initial Inoculum and the Infection Rate in 58 Outbreaks of *Phytophthora infestans* in Maincrop Potato Fields[a]

Initial inoculum	Infection rate: High	Medium	Low	Total
High	8 (3.793)[b]	10 (9.655)	2 (6.552)	20
Medium	3 (3.793)	10 (9.655)	7 (6.552)	20
Low	0 (3.414)	8 (8.690)	10 (5.896)	18
Total	11	28	19	58

[a] From the assessments of Croxall and Smith (1976).

[b] The numbers in parentheses are the numbers calculated to occur if the distribution were random. $\chi^2 = 14.381$; $P < 0.01$.

to the initial inoculum. They tend to be widely, if not randomly, dispersed during planting. The disease to which they give rise tends to be confined in foci; aerial photographs show this. Therefore, disease at any given level in the early part of an epidemic tends to be more widely scattered when it has developed from a high amount of initial inoculum than when it has developed from a low amount. In other words, we associate high initial inoculum with highly scattered disease early on, and, in turn, this high scattering with the high infection rate.

Bibliography

Allen, R. F. (1926). A cytological study of *Puccinia recondita* physiological form 11 on Little Club wheat. *J. Agric. Res.* **33,** 201–222.

Analytis, S. (1973). Zur Methode der Analyse von Epidemien dargestellt am Apfelschorf (*Venturia inaequalis* (Cooke) Aderh.). *Acta Phytomedica* **1,** 1–79.

Anderson, A. J., and Albersheim, P. (1972). Host-pathogen interactions. V. Comparisons of the abilities of proteins isolated from three varieties of *Phaseolus vulgaris* to inhibit the endopolygalacturonases secreted by three races of *Colletotrichum lindemuthianum*. *Physiol. Plant Pathol.* **2,** 339–346.

Anonymous (1954). Verslag van de enquete over het optreden van de aartappelziekte, *Phytophthora infestans* (Mont.) de Bary, in 1953. *Jaarb. Plantenziektenkundig Dienst. Wageningen,* pp. 34–53.

Antonelli, E., and Daly, J. M. (1966). Decarboxylation of indole acetic acid by near-isogenic lines of wheat resistant or susceptible to *Puccinia graminis* f. *sp. tritici*. *Phytopathology* **56,** 610–618.

Asai, G. N. (1960). Intra- and interregional movement of uredospores of black stem rust in the Upper Mississippi River Valley. *Phytopathology* **50,** 535–541.

Aust, H. J., and Kranz, J. (1974). Einfluss der Konidiendichte auf Keimung, Infektion, Inkubationszeit and Sporulation bei dem echten Mehltau der Gerste (*Erysiphe graminis* f. sp. *hordei* Marchal). *Phytopathol. Z.,* **80,** 41–53.

Aust, H. J., Bashi, E., and Rotem, J. (1980). Flexibility of plant pathogens in exploiting ecological and biotic conditions in the development of epidemics. *In* ''Comparative Epidemiology: A Tool for Better Disease Management'' (J. Palti and J. Kranz, eds), pp. 46–56. Center for Agricultural Publishing and Documentation, Wageningen, The Netherlands.

Aylor, D. E. (1978). Dispersal in time and space: Aerial pathogens. *In* "Plant Disease" (J. G. Horsfall and E. B. Cowling, eds.), pp. 159–180. Academic Press, New York.

Barrett, J. A., and Wolfe, M. S. (1978). Multilines and super-races—a reply. *Phytopathology* **68,** 1535–1537.

Bartos, P., Dyk, P. L., and Samborski, D. J. (1969). Adult-plant leaf rust resistance in Thatcher and Marquis wheat: a genetic analysis of the host-parasite situation *Can. J. Bot.* **47,** 267–269.

Beaver, R. G., and Powelson, R. L. (1969). A new race of stripe rust pathogenic on the wheat variety Moro, C.I 13740. *Plant Dis. Rep.* **53,** 91–93.

Berger, R. D. (1977). Application of epidemiological principles to achieve plant disease control. *Annu. Rev. Phytopathol.* **15,** 165–183.

Bhuvaneswari, T. V., Pueppke, S. G., and Bauer, W. D. (1977). Role of lectins in plant-microorganism interactions. I. Binding of soybean lectin to rhizobia. *Plant Physiol.* **60,** 486–491.

Biffen, R. H. (1905). Mendel's laws of inheritance and wheat breeding. *J. Agric. Sci.* **1,** 4–48.

Black, W. (1960). Races of *Phytophthora infestans* and resistance problems in potatoes. *Scot. Plant Breeding Sta. Annu. Rep.* pp. 29–38.

Black, W., Mastenbroek, C., Mills, W. R., and Petersen, L. C. (1953). A proposal for an international nomenclature of races of *Phytophthora infestans* and of genes controlling immunity in *Solanum demissum* derivatives. *Euphytica* **2,** 173–178.

Bohlool, B. B., and Schmidt, E. L. (1974). Lectins: A possible basis for specificity in *Rhizobium*-legume symbiosis. *Science* (*Washington, D.C.*) **185,** 269–271.

Borlaug, N. E. (1946). *Puccinia sorghi* on corn in Mexico. *Phytopathology* **36,** 395.

Borlaug, N. E. (1953). New approach to the breeding of wheat varieties resistant to *Puccinia graminis tritici*. *Phytopathology* **43,** 467.

Borlaug, N. E. (1965). Wheat, rust, and people. *Phytopathology* **55,** 1088–1098.

Bose, A., and Shaw, M. (1974). Growth of rust fungi of wheat and flax on chemically defined media. *Nature* (*London*) **251,** 646–648.

Bracker, C. E., and Littlefield, L. J. (1973). Structural concepts of host-pathogen interfaces. *In* "Fungal Pathogenicity and the Plant's Response" (R. J. W. Byrde and C. V. Cutting, eds.), pp. 159–315. Academic Press, New York.

Brinkerhoff, L. A. (1963). Variability of *Xanthomonas malvacearum,* the cotton blight organism. *Okla. Agric. Exp. Stn. Tech. Bull.* **T98,** 1–96.

Brinkerhoff, L. A. (1970). Variation in *Xanthomonas malvacearum* and its relation to control. *Annu. Rev. Phytopathol.* **8,** 85–110.

Brinkerhoff, L. A., and Presley, J. T. (1967). Effect of four day and night temperature regimes on bacterial blight of immune, resistant and susceptible strains of upland cotton. *Phytopathology* **57,** 47–51.

Bromfield, K. R. (1961). The effect of postinoculation temperature on seedling reaction of selected wheat varieties to stem rust. *Phytopathology* **51,** 590–593.

Caldwell, R. M., Roberts, J. J., and Eyal, Z. (1970). General resistance ("slow rusting") to *Puccinia recondita* f. *sp. tritici* in winter and spring wheats. *Phytopathology* **60,** 1287.

Caten, C. E. (1970). Spontaneous variability of single isolates of *Phytophthora infestans*. II. Pathogenic variation. *Can. J. Bot.* **48,** 897–905.

Chakravorty, A. K., and Shaw, M. (1977). A possible molecular basis for obligate host-pathogen interactions. *Biol. Rev.* **52,** 147–149.

Chen, A. P., and Phillips, D. A. (1976). Attachment of Rhizobium to legume roots as the basis of specific interactions. *Physiol. Plant* **38,** 83–88.

Cirulli, M., and Alexander, L. J. (1966). A comparison of pathogenic isolates of *Fusarium oxysporum f. lycopersici* and different sources of resistance in tomato. *Phytopathology* **56,** 1301–1304.

Cirulli, M., and Ciccarese, F. (1975). Interactions between TMV isolates, temperature, allelic condition and combination of the *Tm* resistance genes in tomato *Phytopathol. Mediterr.* **14,** 100–105.

Clifford, B. C., and Clothier, R. B. (1974). Physiologic specialization of *Puccinia hordei* on barley hosts with non-hypersensitive resistance. *Trans. Br. Mycol. Soc.* **63,** 421–430.

Crill, P. (1977). An assessment of stabilizing selection in crop variety development. *Annu. Rev. Phytopathol.* **15,** 185–202.

Cross, J. E. (1963). Pathogenic differences in Tanganyika populations of *Xanthomonas malvacearum*. *Emp. Cotton Grow. Rev.* **40,** 125–130.

Crosse, J. E. (1975). Variation amongst plant pathogenic bacteria. *Ann. Appl. Biol.* **81,** 834.

Croxall, H. E., and Smith, L. P. (1976). The epidemiology of potato blight in the East Midlands 1923–1974. *Ann. Appl. Biol.* 82, 451–466.

Daly, J. M. (1949). The influence of nitrogen source on the development of stem rust of wheat. *Phytopathology* **39,** 386–393.

Daly, J. M. (1972). The use of near-isogenic lines in biochemical studies of the resistance of wheat to stem rust. *Phytopathology* **62,** 392–400.

Daly, J. M., Ludden, P., and Seevers, P. M. (1974). Biochemical comparisons of resistance to wheat stem rust disease controlled by the *Sr6* and *Sr11* alleles. *Physiol. Plant Pathol.* **1,** 397–407.

Day, P. R. (1974). "Genetics of Host–Parasite Interaction." Freeman, San Francisco, California.

Dazzo, F. B., and Hubbell, D. H. (1975). Cross-reactive antigens and lectin as determinants of symbiotic specificity in the Rhizobium-clover association. *Appl. Microbiol.* **30,** 1017–1033.

Dazzo, F. B., Napoli, C. A., and Hubbell, D. H.. (1976). Adsorption of bacteria to roots as related to host specificity in the *Rhizobium*-clover symbiosis. *Appl. Environ. Microbiol.* **32,** 166–171.

de Bary, A. (1876). Researches into the nature of the potato fungus (*Phytophthora infestans*). *J. R. Agric. Soc. Engl.* **12,** 239.

de Bruijn, H. L. G. (1951). Pathogenic differentiation in *Phytophthora infestans* (Mont.) de Bary. *Phytopathol. Z.* **18,** 339–359.

Denward, T. (1970). Differentiation in *Phytophthora infestans*. II. Somatic recombination in vegetative mycelium. *Heriditas* **66,** 35–48.

De Vay, J. E., and Adler, H. E. (1976). Antigens common to hosts and parasites. *Annu. Rev. Microbiol.* **30,** 147–168.

De Vay, J. E., Schnathorst, W. C., and Foda, M. S. (1967). Common antigens and host-parasite interactions. *In* "The Dynamic Role of Molecular Constituents in Plant-parasitic Interactions" (C. Mirocha and L. Uritani, eds.,) pp. 313–328. Bruce Publ., St. Paul, Minnesota.

Dobzhansky, T. (1943). Genetics of natural populations. IX. Temporal changes in the populations of *Drosophila pseudoobscura*. *Genetics* **28,** 162–186.

Doke, N., and Tomiyama, K. (1977). Effect of high molecular weight substances released from zoospores of *Phytophthora infestans* on hypersensitive response of potato tubers. *Phytopathol. Z.* **90,** 236–242.

Doke, N., Garas, N. A., and Kuć, J. (1977). Partial characterization and mode of action of the hypersensitivity-inhibiting factor isolated from *Phytophthora infestans*. *Proc. Am. Phytopathol. Soc.* **4,** 165.

Doke, N., Garas, N. A., and Kuć, J. (1979). Partial characterization and aspects of the mode of action of a hypersensitivity-inhibiting factor (HIF) isolated from *Phytophthora infestans*. *Physiol. Plant Pathol.* **15,** 169–175.

Doke, N., Garas, N. A., and Kuć, J. (1980). Effect on host hypersensitivity of suppressors released during the germination of *Phytophthora infestans* cystospores. *Phytopathology* **70,** 35–39.

Doubly, J. A., Flor, H. H., and Clagett, C. O. (1960). Relation of antigens of *Melampsora lini* and *Linum usitatissimum* to resistance and susceptibility. *Science (Washington, D.C.)* **131,** 229.

Dyck, P. L., and Samborski, D. J. (1970). The genetics of two alleles for leaf rust resistance at the *Lr*14 locus in wheat. *Can. J. Genet. Cytol.* **12,** 689.

Ebba, T., and Person, C. (1975). Genetic control of virulence in *Ustilago hordei*. IV. Duplicate genes for virulence and genetic and environmental modification of a gene-for-gene relationship. *Can. J. Genet. Cytol.* **17,** 631–636.

Ebel, J., Ayers, A. R., and Albersheim, P. (1976). Host-pathogen interactions. XII. Response of suspension-cultured soybean cells to the elicitor isolated from *Phytophthora megasperma* var. *sojae,* a fungal pathogen of soybeans. *Plant Physiol.* **57,** 775–779.

Eide, C. J., Bonde, R., Gallegly, M. E., Graham, K. M., Mills, W. R. Niederhauser, J., and Wallin, J. R. (1959). Report of the late blight investigations committee. *Am. Potato J.* **36,** 421–423.

Ellingboe, A. H. (1976). Genetics of host-parasite interactions. *In* "Physiological Plant Pathology" (R. Heitefuss and P. H. Williams, eds.), Encyclopedia of Plant Physiology, Vol. 4, pp. 760–778. Springer-Verlag, Berlin and New York.

Elliott, C., and Jenkins, M. T. (1946). Helminthosporium leaf blight of corn. *Phytopathology* **36,** 660–666.

Estrada, R. N., and Guzman, N. (1969). Herencia de la resistencia de campo al "tizon" (*Phytophthora infestans* Mont. de Bary). En variedades cultivados de papa (subspecies *tuberosa y andigena*). *Rev. Inst. Colomb. Agropecic* **4,** 117–137.

Farkas, G. L. (1978). Senescence and plant disease. *In* "Plant Disease" (J. G. Horsfall and E. B. Cowling, eds.), Vol. III, pp. 391–412. Academic Press, New York.

Fedotova, T. I. (1948). Significance of individual proteins of seed in the manifestation of the resistance of plants to diseases. *Tr. Leningr. Inst. Zasch. Rast. Sb.* **1,** 61–71.

Finlay, K. W. (1952). Inheritance of spotted wilt resistance in the tomato. I. Identification of strains of the virus by the resistance or susceptibility of tomato species. *Aust. J. Sci. Res. Ser. B.* **5,** 303–314.

Fisher, R. A. (1930). "The Genetical Theory of Natural Selection." Oxford Univ. Press (Clarendon), London and New York.

Fleming, R. A. (1980). Selection pressures and plant pathogens: Robustress of the model. *Phytopathology* **70,** 175–178.

Flor, H. H. (1942). Inheritance of pathogenicity of *Melampsora lini*. *Phytopathology* **32,** 653–669.

Flor, H. H. (1953). Epidemiology of flax rust in the North Central States. *Phytopathology* **43,** 624–628.

Flor, H. H. (1956). The complementary genic systems in flax and flax rust. *Adv. Gene.* **8,** 29–54.

Flor, H. H. (1958). Mutation to wider virulence in *Melampsora lini*. *Phytopathology* **48,** 297–301.

Flor, H. H. (1960). The inheritance of X-ray induced mutations to virulence in a uredospore culture of race 1 of *Melampsora lini*. *Phytopathology* **50,** 603–605.

Flor, H. H. (1965). Tests for allelism of rust resistance genes in flax. *Crop Sci.* **5,** 415–418.

Flor, H. H. (1971). Current status of the gene-for-gene concept. *Annu. Rev. Phytopathol.* **9,** 275–296.

Frandsen, N. D. (1956). Rasse 4 von *Phytophthora infestans* in Deutschland. *Phytopathol. Z.* **26,** 124–130.

Frey, K. J., Browning, J. A., and Simons, M. D. (1977). Management systems for host genes to control disease loss. *Ann. N. Y. Acad. Sci.* **287,** 255–274.

Fuchs, E., and Hille, M. (1968). The behaviour of some yellow rust races on differential varieties under different environmental conditions. *Rep. Cereal Rusts Conf. Oeiras Portugal* 146–151.

Futrell, M. C., and Dickson, J. G. (1954). The influence of temperature on the development of powdery mildew of spring wheats. *Phytopathology* **44,** 247–251.

Gabe, H. L. (1975). Standardization of nomenclature for pathogenic races of *Fusarium oxysporum* f. sp. *lycopersici*. *Trans. Br. Mycol. Soc.* **64,** 156–159.

Gäumann, E. (1946). "Pflanzliche Infektionslehre." Birkhaeuser, Basel.

Gassner, G., and Kirchhoff, H. (1934). Einige vergleichende Versuche über Verschiebungen der Rostresistenz in Abhänzigkeit vom Entwicklungszustand der Getreidepflanzen. *Phytopathol. Z.* **7,** 43-52.

Gassner, G., and Straib, W. (1932). Über Mutationen in einer biologischen Rasse von *Puccinia glumarum tritici* (Schmidt) Erikss. und Henn. *Z. Indukt. Abstammungs—und Vererbungslehre* **43,** 155-180.

Gassner, G., and Straib, W. (1934). Experimentelle Untersuchungen zur Epidemiologie des Gelbrostes (*Puccinia glumarum* (Schm.) Erikss. and Henn.) *Phytopathol. Z.* **7,** 285-307.

Gevers, H. A. (1975). A new major gene for resistance to *Helminthosporium turcicum* leaf blight of maize. *Plant Dis. Rep.* **59,** 296-299.

Golik, I. V., Gromova, B. B-O., and Fedotova, T. I. (1977). Immunochemical similarities of proteins of *Synchytrium endobioticum* (Schilb.) Perc. and the host plant. Abst. in *Rev. Plant Pathol.* **56,** 84-85.

Goodman, R. N., Király, Z., and Zaitlin, M. (1967). "The Biochemistry and Physiology of Infectious Plant Disease." Van Nostrand-Reinhold, Princeton, New Jersey.

Gordon, W. L. (1930). Effect of temperature on host reactions to physiologic forms of *Puccinia graminis avenae* Erikss. and Henn. *Sci. Agric.* **11,** 95-103.

Gordon, W. L. (1933). A study of the relation of environment to the development of the uredinial and telial stages of the physiologic forms of *Puccinia graminis avenae Erikss. and Henn. Sci. Agric.* **14,** 184-237.

Gough, F. J., and Merkle, O. G. (1971). Inheritance of stem and leaf rust resistance in Agent and Agrus cultivars of *Triticum aestivum. Phytopathology* **61,** 1501-1505.

Graf-Marin, A. (1934). Studies on powdery mildew of cereals. *Cornell Agr. Exp. Sta. Memoir* **157,** 48 pp.

Graham, K. M. (1955). Distribution of physiological races of *Phytophthora infestans* (Mont.) de Bary in Canada, *Am. Potato J.* **32,** 277-288.

Graham, K. M., Dionne, L. A., and Hodgson, W. A. (1961). Mobility of *Phytophthora infestans* on blight resistant selections of potatoes and tomatoes. *Phytopathology* **51,** 264-265.

Grainger, J. (1956). Host nutrition and attack by fungal parasites. *Phytopathology* **46,** 445-456.

Grainger, J. (1957). Blight—the potato versus *Phytophthora infestans. Agric. Rev.* **3,** 10-26.

Grainger, J. (1959). Effects of diseases on crop plants. *Outlook Agric.* **2,** 114-121.

Green, G. J. (1971a). Stem rust of wheat, barley and rye in Canada in 1970. *Can. Plant Dis. Surv.* **51,** 20-23.

Green, G. J. (1971b). Physiologic races of wheat stem rust in Canada from 1919 to 1969. *Can. J. Bot.* **49,** 1575-1588.

Green, G. J. (1972a). Stem rust of wheat, barley and rye in Canada in 1971. *Can. Plant Dis. Surv.* **52,** 11-14.

Green, G. J. (1972b). Stem rust of wheat, barley and rye in Canada in 1972. *Can. Plant Dis. Surv.* **52,** 162-167.

Green, G. J. (1974). Stem rust of wheat, barley and rye in Canada in 1973. *Can. Plant Dis. Surv.* **54,** 11-16.

Green, G. J. (1975). Stem rust of wheat, barley and rye in Canada in 1974. *Can. Plant Dis. Surv.* **55,** 51-57.

Green, G. J. (1976a). Stem rust of wheat, barley and rye in Canada in 1975. *Can. Plant Dis. Surv.* **56,** 15-18.

Green, G. J. (1976b). Stem rust of wheat, barley and rye in Canada in 1976. *Can. Plant Dis. Surv.* **56,** 119-122.

Green, G. J. (1976c). Axenic culture of *Puccinia* species collected in Canada. *Can. J. Bot.* **54,** 1198-1205.

Green, G. J. (1978). Stem rust of wheat, barley and rye in Canada in 1977. *Can. Plant Dis. Surv.* **58,** 44–48.

Green, G. J. (1979). Stem rust of wheat, barley and rye in Canada in 1978. *Can. Plant Dis. Surv.* **59,** 43–47.

Green, G. J., and Campbell, A. B. (1979). Wheat cultivars resistant to *Puccinia graminis tritici* in western Canada: their development, performance, and economic value. *Can. J. Plant Pathol.* **1,** 3–11.

Green, G. I., Knott, D. R., Watson, I. A., and Pugsley, A. T. (1960). Seedling reactions to stem rust of lines of Marquis wheat with substituted genes for rust reaction. *Can. J. Plant Sci.* **40,** 524–538.

Gregory, P. H. (1945). The dispersion of air-borne spores. *Trans. Br. Mycol. Soc.* **28,** 26–72.

Grogan, C. O., and Rosenkranz, E. E. (1968). Genetics of host reaction to corn stunt virus. *Crop Sci.* **8,** 251–254.

Groth, J. V. (1976). Multilines and "super races"; a simple model. *Phytopathology* **66,** 937–939.

Groth, J. V. (1978). Rebuttal to "multilines and super-races—a reply." *Phytopathology* **68,** 1538–1539.

Guseva, N. N., and Gromova, B. B-O. (1976). Cross-reacting antigens of the brown rust causal agent and wheat plants in immunological tests. *In* "Proceedings of the Fourth European and Mediterranean Cereal Rusts Conference, Interlaken, Switzerland 1976" (A. Bronnimann, ed.), p. 13. European and Mediterranean Cereal Rusts Foundation, Interlaken.

Hadidi, A., and Fraenkel-Conrat, H. (1973). Characterization and specificity of soluble RNA polymerase of brome mosaic virus. *Virology* **52,** 363–372.

Hanchey, P., and Wheeler, H. (1971). Pathological changes in ultrastructure: Tobacco roots infected with *Phytophthora parasitice* var. *nicotianae*. *Phytopathology* **61,** 33–39.

Hariharasubramanian, V., Hadidi, A., Singer, B., and Fraenkel-Conrat, H. (1973). Possible verification of a protein in brome mosaic virus infected barley as a compound of viral RNA replicase. *Virology* **54,** 190–198.

Harrison, B. D. (1977). Ecology and control of viruses with soil-inhabiting vectors. *Annu. Rev. Phytopathol.* **15,** 331–360.

Hart, H. (1949). Nature and variability of disease resistance in plants. *Annu. Rev. Microbiol.* **3,** 289–316.

Harvey, A. E., and Grasham, J. L. (1974). Axenic culture of the mononucleate stage of *Cronartium ribicola*. *Phytopathology* **64,** 1028–1035.

Henderson, S. J., and Friend, J. (1979). Increase in PAL and lignin-like compounds as race-specific responses of potato tubers to *Phytophthora infestans*. *Phytopathol. Z.* **94,** 323–334.

Hingorani, M. K. (1947). Factors affecting the survival ability of certain physiological races of *Puccinia graminis avenae* E. and H. Ph.D. thesis, Univ. of Minnesota, Minneapolis (Quoted by Hart, 1949).

Hoff, R. J., and McDonald, G. I. (1980). Resistance to *Cronartium ribicola* in *Pinus monticola:* reduced needle-spot frequency. *Can. J. Bot.* **58,** 574–577.

Hollis, C. A., Schmidt, R. A., and Kimbrough, J. W. (1972). Axenic culture of *Cronartium fusiforme*. *Phytopathology* **62,** 1417–1419.

Hooker, A. L. (1962a). Additional sources of resistance to *Puccinia sorghi* in the United States. *Plant Dis. Rep.* **46,** 14–16.

Hooker, A. L. (1962b). Corn leaf disease. *Proc. 17th Annu. Hybrid Corn Indust. Res. Conf.* **17,** 24–36.

Hooker, A. L. (1963). Monogenic resistance in *Zea mays* L. to *Helminthosporium turcicum*. *Crop Sci.* **3,** 381–383.

Hooker, A. L. (1973). New developments in the corn leaf and stalk disease picture. *Proc. 28th Annu. Hybrid Corn and Sorghum Conf.* **28,** 62–71.

Hooker, A. L. (1977). A second major gene locus in corn for chlorotic lesion resistance to *Helminthosporium turcicum*. *Crop. Sci.* **17,** 132–135.
Hooker, A. L. (1979). Breeding for resistance to some complex disease of corn. *In* "International Rice Research Institute Rice Blast Workshop" pp. 153–181. Los Banos, Laguna, Philippines.
Hooker, A. L., and Le Roux, P. M. (1957). Sources of protoplasmic resistance to *Puccinia sorghi* in corn. *Phytopathology* **47,** 187–191.
Hooker, A. L., and Tsung, Y-K (1980). Relationship of dominant genes in corn for chlorotic lesion resistance to *Helminthosporium turcicum*. *Plant Dis.* **64,** 387–388.
Hooker, A. L., Sprague, G. F., and Russell, W. A. (1955). Resistance to rust (*Puccinia sorghi*) in corn. *Agron. J.* **47,** 388.
Howatt, J. L., and Grainger, P. N. (1955). Some new findings concerning *Phytophthora infestans* (Mont.) de By. *Am. Potato J.* **32,** 180–188.
Hoyle, M. C. (1977). High resolution of peroxidase-indoleacetic acid oxidase enzymes from horseradish by isoelectric focussing. *Plant Physiol.* **60,** 787–793.
Hughes, G. R., and Hooker, A. L. (1971). Genes conditioning resistance to northern leaf blight in maize. *Crop Sci.* **11,** 180–184.
Ibrahim, I. (1949). Effect of some conditions and chemicals on the development of races 2, 6, 7, and 8 of *Puccinia graminis avenae* E. and H. Master's thesis, Univ. of Minnesota, Minneapolis (Quoted by Hart, 1949).
Jenkins, M. T., Robert, A. L., and Findley, W. R. (1954). Recurrent selection as a method for concentrating genes for resistance to *Helminthosporium turcicum* leaf blight in corn. *Agron. J.* **46,** 89–94.
Jenkyn, J. F. (1973). Seasonal changes in incubation time in *Erysiphe graminis* f. sp. *hordei*. *Ann. Appl. Biol.* **73,** 15–18.
Jensen, N. F. (1952). Intra-varietal diversification in oat breeding. *Agron. J.* **44,** 30–34.
Johnson, D. A., and Wilcoxson, R. D. (1978). Components of slow rusting in barley infected with *Puccinia hordei*. *Phytopathology* **68,** 1470–1474.
Johnson, L. B., and Schafer, J. F. (1965). Identification of wheat leaf rust resistance combinations by differential temperature effects. *Plant Dis. Rep.* **49,** 222–224.
Johnson, R., and Taylor, A. J. (1976). Spore yield of pathogens in investigations of race-specificity of host resistance. *Annu. Rev. Phytopathol.* **14,** 97–119.
Johnson, T. (1931). Studies in cereal diseases. VI. A study of the effect of environmental factors on the variability of physiologic forms of *Puccinia graminis tritici* Erikss. and Henn. *Can. Dept. Agric. Bull.* **140,** *NS.* 76 pp.
Jones, D. R. (1973). Growth and nutritional studies with axenic cultures of the carnation rust fungus, *Uromyces dianthi* (Pers.) Niessl. *Physiol. Plant Pathol.* **3,** 379–386.
Jones, D. R., and Deverall, B. J. (1977). The effect of the Lr20 resistance gene in wheat on the development of leaf rust, *Puccinia recondita*. *Physiol. Plant Pathol.* **10,** 275–285.
Jones, E. D., and Mullen, J. M. (1974). The effect of potato virus X on susceptibility of potato tubers to *Fusarium roseum* "Avenaceum". *Am. Potato J.* **51,** 209–215.
Jones, I. T., and Hayes, J. D. (1971). The effect of sowing date on adult plant resistance to *Erysiphe graminis* f. sp. *avenue* in oats. *Ann. Appl. Biol.* **68,** 31–39.
Kamen, R. (1970). Characterization of the subunits of Qβ replicase. *Nature* (*London*) **228,** 527–533.
Kappelman, A. J., and Thompson, D. L. (1966). Inheritance of resistance to Diplodia stalk-rot in corn. *Crop Sci.* **6,** 288–290.
Karanmal, A., Watkins, J. E., and Dunkle, L. D. (1978). Virulence distribution of *Puccinia recondita* in Nebraska in 1977. *Phytopathol. News* **12,** 89.
Katsuya, K., and Green, G. J. (1967). Reproductive potentials of races 15B and 56 of wheat stem rust. *Can. J. Bot.* **45,** 1077–1091.
Keegstra, K., Talmadge, K. W., Bauer, W. D., and Albersheim, P. (1973). The structure of plant cell walls. *Plant Physiol.* **51,** 188–196.

Kim, S. K., and Brewbaker, J. L. (1977). Inheritance of general resistance in maize to *Puccinia sorghi*. *Crop Sci.* **17,** 456–461.

Klement, Z., and Goodman, R. N. (1967). The hypersensitivity reaction to infection by bacterial plant pathogens. *Annu. Rev. Phytopathol.* **5,** 17–44.

Knott, D. R., and Srivastava, J. P. (1977). Inheritance of resistance to stem rust races 15B and 56 in eight cultivars of common wheat. *Can. J. Plant Sci.* **57,** 633–641.

Kondo, M., Gallerani, R., and Weissmann, C. (1970). Subunit structure of Qβ replicase. *Nature (London)* **228,** 525–527.

Kranz, J. (1974). The role and scope of mathematical analysis and modeling in epidemiology. *In* "Epidemics of Plant Diseases: Mathematical Analysis and Modeling" (J. Kranz, ed.), pp. 7–54. Springer-Verlag, Berlin and New York.

Kranz, J. (1975). Das Abklingen von Befallskurven. *Z. Pflanzenkrankheiten u. Pflanzenschutz* **82,** 655–664.

Kranz, J. (1977). A study of maximum severity in plant diseases. *In* "Travaux Dédiés à G. Viennot-Bourgin," pp. 169–173. Société Française de Phytopathologie, Paris.

Krause, R. A., Massie, L. B., and Hyre, R. A. (1975). Blitecast: a computerized forecast of potato late blight. *Plant Dis. Rep.* **59,** 95–98.

Krupinsky, J. M., and Sharp, E. L. (1979). Reselection for improved resistance of wheat to stripe rust. *Phytopathology* **69,** 400–404.

Lauffer, M. A. (1975). "Entropy—Driven Processes in Biology." Springer-Verlag, Berlin and New York.

Leonard, K. J. (1977). Selection pressures and plant pathogens: genetic basis of epidemics in agriculture. *Ann. N. Y. Acad. Sci.* **287,** 202–222.

Leonard, K. J., and Czochor, R. J. (1978). In response to "Selection pressures and plant pathogens: stability of equilibria". *Phytopathology* **68,** 971–973.

Leonard, K. J., and Czochor, R. J. (1980). Theory of genetic interactions among populations of plants and their pathogens. *Annu. Rev. Phytopathol.* **18,** 237–258.

Lie, T. A., Hille, D., Lambers, R., and Houwers, A. (1976). Symbiotic specialization in pea plants: some environmental effects on nodulation and nitrogen fixation. *In* "Symbiotic Nitrogen Fixation in Plants" (P. S. Nutman, ed.), pp. 319–333. Cambridge Univ. Press, London and New York.

Lim, S. M. (1975). Diallel analysis for reaction of eight corn hybrids to *Helminthosporium maydis* race *T. Phytopathology* **65,** 10–15.

Lim, S. M., Kinsey, J. G., and Hooker, A. L. (1974). Inheritance of virulence in *Helminthosporium turcicum* to monogenic resistant corn. *Phytopathology* **64,** 1150–1151.

Lin, M-R., and Edwards, H. N. (1974). Primary penetration process in powdery mildewed barley related to host cell age, cell type, and occurrence of basic staining material. *New Phytol,* **73,** 131–137.

Lipetz, J. (1970). Wound healing in higher plants. *Int. Rev. Cytol.* **27,** 1–28.

Lippincott, B. B., and Lippincott, J. A. (1969). Bacterial attachment to a specific wound site as an essential stage in tumor initiation by *Agrobacterium tumefaciens*. *J. Bacteriol.* **97,** 620–628.

Lippincott, B. B., Whatley, M. H., and Lippincott, J. A. (1977). Tumor induction by *Agrobacterium* involves attachment of the bacterium to a site on the host cell wall. *Plant Physiol.* **59,** 388–390.

Lippincott, J. A., and Lippincott, B. B. (1977). Nature and specificity of the bacterium-host attachment in *Agrobacterium* infection. *In* "Cell Wall Biochemistry Related to Specificity in Host–Pathogen Interactions" (B. Solheim and J. Raa, eds.), pp. 239–251. Universiteitsforlaget, Oslo.

Ludden, P., and Daly, J. M. (1970). Certain biochemical comparisons in wheat stem rust disease reactions controlled either at the *Sr*6 or *Sr*11 loci. *Phytopathology* **60,** 1301.

Luig, N. H. (1979). Mutation studies in *Puccinia graminis tritici*. *Proc. 5th Intern. Wheat Genetics Symposium,* 533–539.

Luig, N. H., and Baker, E. P. (1973). Variability in oat stem rust in Eastern Australia. *Proc. Linnean Soc. N. S. Wales* **98,** 53–61.

Luig, N. H., and Rajaram, S. (1972). The effect of temperature and genetic background on the host gene expression and interaction with *Puccinia graminis tritici*. *Phytopathology* **62,** 1171–1174.

Luig, N. H., and Watson, I. A. (1965). Studies on the genetic nature of resistance to *Puccinia graminis* var. *tritici* in six varieties of common wheat. *Proc. Linnean Soc. N.S. Wales* **90,** 299–327.

Luig, N. H., and Watson, I. A. (1970). The effect of complex genetic resistance in wheat on the variability of *Puccinia graminis* f. sp. *tritici*. *Proc. Linnean Soc. N.S. Wales* **95,** 22–45.

Macko, V., Woodbury, W., and Stahmann, M. A. (1968). The effect of peroxidase on the germination and growth of mycelium of *Puccinia graminis* f. sp. *tritici*. *Phytopathology* **58,** 1250–1254.

Mahomet, H. A. (1954). Temperature requirements for the identification of races 49 and 139 of *Puccinia graminis tritici*. *Phytopathology* **44,** 498.

Malcolmson, J. F. (1969). Races of *Phytophthora infestans* occurring in Great Britain. *Trans. Br. Mycol. Soc.* **53,** 417–423.

Martens, J. W. (1972). Stem rust of oats in Canada in 1972. *Can. Plant Dis. Surv.* **52,** 171–172.

Martens, J. W. (1974). Stem rust of oats in Canada in 1973. *Can. Plant Dis. Surv.* **54,** 19–20.

Martens, J. W. (1975). Stem rust of oats in Canada in 1974. *Can. Plant Dis. Surv.* **55,** 61–62.

Martens, J. W. (1978). Stem rust of oats in Canada in 1977. *Can. Plant Dis. Surv.* **58,** 51–52.

Martens, J. W. (1979). Stem rust of oats in Canada in 1978. *Can. Plant Dis. Surv.* **59,** 70–72.

Martens, J. W., and Anema, P. K. (1972). Stem rust of oats in Canada in 1971. *Can. Plant Dis. Surv.* **52,** 17–18.

Martens, J. W., and McKenzie, R. I. H. (1976a). Stem rust of oats in Canada in 1975. *Can. Plant Dis. Surv.* **56,** 23–24.

Martens, J. W., and McKenzie, R. I. H. (1976b). Stem rust of oats in Canada in 1976. *Can. Plant Dis. Surv.* **56,** 126–128.

Martens, J. W., McKenzie, R. I. H., and Green, G. J. (1967). Thermal stability of stem rust resistance in oat seedlings. *Can. J. Bot.* **45,** 451–458.

Martens, J. W., McKenzie, R. I. H., and Green, G. J. (1970). Gene-for-gene relationships in the *Avena—Puccinia graminis* host-parasite system in Canada. *Can. J. Bot.* **48,** 969–975.

Martens, J. W., Green, G. J., and McKenzie, R. I. H. (1976). International oat stem rust virulence survey. *Plant Dis. Rep.* **60,** 525–528.

Martens, J. W., Burnett, P. A., and Wright, G. M. (1977). Virulence in *Puccinia coronata* f.sp. *avenae* and *P. graminis* f.sp. *avenae* in New Zealand. *Phytopathology* **67,** 1519–1521.

Martens, J. W., Roelfs, A. P., McKenzie, R. I. H., Rothman, P. G., Stuthman, D., and Brown, P. D. (1979). Systems of nomenclature for races of *Puccinia graminis* f. sp. *avenae*. *Phytopathology* **69,** 293–294.

Mather, K. (1943). Polygenic inheritance and natural selection. *Biol. Rev.* **18,** 32–64.

McIntosh, R. A. (1973). A catalogue of gene symbols for wheat. *Proc. 4th Int. Wheat Genet. Symp. Missouri Agr. Exp. Sta.,* Columbia, Mo. pp. 893–937.

McIntosh, R. A. (1977). Nature of induced mutations affecting disease reactions in wheat. *In* "Induced Mutations against Plant Diseases," pp. 551–565. Intern. Atomic Energy Agency, Vienna.

McIntosh, R. A., Luig, N. H., and Baker, E. P. (1967). Genetic and cytogenic studies of stem rust, leaf rust, and powdery mildew resistances in Hope and related wheat cultivars. *Aust. J. Biol. Sci.* **20,** 1181–1192.

Mendgen, K. (1975). Ultrastructural demonstration of different peroxidase activities during the bean rust infection process. *Physiol. Plant Pathol.* **6,** 275–282.

Mercer, P. C., Wood. R. K. S., and Greenwood, A. D. (1974). Resistance to anthracnose of French bean. *Physiol. Plant Pathol.* **4,** 291–306.

Mercer, P. C., Wood, R. K. S., and Greenwood, A. D. (1975). Ultrastructure of the parasitism of *Phaseolus vulgaris* by *Colletotrichum lindemunthianum. Physiol. Plant Pathol.* **5,** 203–214.

Mickle, J. (1845). Potato murrain. *Gard. Chron.* **5,** 658.

Miles, J. W., Dudley, J. W., White, D. G., and Lambert, R. J. (1980). Improving corn populations for grain yield and resistance to leaf blight and stalk rot. *Crop Sci.* **20,** 247–251.

Milus, E. A., and Line, R. F. (1980). Virulence in *Puccinia recondita* in the Pacific Northwest. *Plant Dis.* **64,** 78–80.

Moll, R. H., Thompson, D. L., and Harvey, P. H. (1963). A quantitative genetic study of the inheritance of resistance to brown spot (*Physoderma maydis*) of corn. *Crop Sci.* **3,** 389–391.

Morris, E. R., Rees, D. A., Young, G., Walkinshaw, M. D., and Darke, A. (1977). Order-disorder transition for a bacterial polysaccharide in solution: A role for polysaccharide conformation in recognition between *Xanthomonas* pathogen and its host plant. *J. Mol. Biol.* **110,** 1–16.

Müller, K. O. (1931). Über die Entwicklung von *Phytophthora infestans* auf anfälligen und widerstandsfähigen Karkoffelsorten. *Arb. Biol. Anst. Dahlem* **18,** 465–505.

Müller, K. O. (1950). Affinity and reactivity of angiosperms to *Phytophthora infestans. Nature (London)* **166,** 392–394.

Nelson, R. R. (1978). Genetics of horizontal resistance to plant diseases. *Annu. Rev. Phytopathol.* **16,** 359–378.

Newton, M., and Johnson, T. (1939). A mutation for pathogenicity in *Puccinia graminis tritici. Res.* C. **22,** 201–216.

Ohm, H. W., and Shaner, G. E. (1976). Three components of slow leaf-rusting at different stages in wheat. *Phytopathology* **66,** 1356–1360.

Oosawa, F., and Asakura, S. (1975). "Thermodynamics of the Polymerization of Protein." Academic Press, New York.

Osora, M. O., and Green, G. J. (1976). Stabilizing selection in *Puccinia graminis tritici* in Canada. *Can. J. Bot.* **54,** 2204–2214.

Palmerley, R. A., and Callow, J. A. (1978). Common antigens in extracts of *Phytophthora infestans* and potatoes. *Physiol. Plant Pathol.* **12,** 241–248.

Parlevliet, J. E. (1976). Partial resistance of barley to leaf rust, *Puccinia hordei*. III. The inheritance of the host plant effect on the latent period of four cultivars. *Euphytica* **25,** 241–248.

Parlevliet, J. E., and Kuiper, H. T. (1977). Resistance of some barley cultivars to leaf rust; polygenic partial resistance hidden by monogenic hypersensitivity. *Neth. J. Plant Pathol.* **83,** 85–89.

Parlevliet, J. E., and Zadoks, J. C. (1977). The integrated concept of disease resistance: a new view including horizontal and vertical resistance in plants. *Euphytica* **26,** 5–21.

Paxman, G. J. (1963). Variation in *Phytophthora infestans. Europ. Potato J.* **6,** 14–23.

Pelham, J. (1966). Resistance in tomato to tobacco mosaic virus. *Euphytica* **15,** 258–267.

Pelham, J., Fletcher, J. T., and Hawkins, J. H. (1970). The establishment of a new strain of tobacco mosaic virus resulting from the use of resistant varieties of tomato. *Ann. Appl. Biol.* **65,** 293–297.

Person, C., and Ebba, T. (1975). Genetics of fungal pathogens. *Genet. Supp.* **79,** 397–408.

Person, C., Groth, J. V., and Mylyk, O. M. (1976). Genetic change in host-parasite populations. *Annu. Rev. Phytopathol.* **14,** 177–188.

Perutz, M. F. (1978). Electrostatic effects in proteins. *Science* **201,** 1187–1191.

Peterson, R. S., and Jewell, F. F. (1968). Status of American stem rusts of pine. *Annu. Rev. Phytopathol.* **6,** 23–40.

Peturson, B. (1930). Effect of temperature on host reactions to physiologic forms of *Puccinia coronata avenae. Sci. Agric.* **11,** 104–110.

Pochard, E., Gaujon, C., and Vergara, S. (1962). Influence de la température, de l'eclairement et du stade de la plante sur l'expression de la sensibilité à un biotype de la rouille jaune de quelques variétiés de blé. *Ann. Amélior. Plantes* **12,** 45–58.

Populer, C. (1978). Changes in host susceptibility with time. *In* "Plant Diseases" (J. G. Horsfall and E. B. Cowling, eds.), Vol. II, pp. 239–262. Academic Press, New York.

Priestley, R. H. (1978). Detection of increased virulence in populations of wheat yellow rust. *In* "Plant Disease Epidemiology" (P. R. Scott and A. Bainbridge, eds.), pp. 63–70. Blackwell, Oxford.

Rajaram, S., Luig, N. H., and Watson, I. A. (1971). Inheritance of leaf rust resistance in four varieties of common wheat. *Euphytica* **20,** 574–585.

Rao, Y. P., Mohan, S. K., and Reddy, P. R. (1971). Pathogenic variability in *Xanthomonas oryzae*. *Plant Dis. Rep.* **55,** 593–595.

Reddick, D., and Mills, W. R. (1938). Building up virulence in *Phytophthora infestans*. *Am. Potato J.* **15,** 29–34.

Reddy, O. R., and Ou, S. H. (1976). Pathogenic variability in *Xanthomonas oryzae*. *Phytopathology* **66,** 906–909.

Rees, R. G., Thompson, J. P., and Goward, E. A. (1979a). Slow rusting and tolerance to rusts in wheat. II. The progress and effects of epidemics of *Puccinia recondita tritici* in selected wheat cultivars. *Aust. J. Agric. Res.* **30,** 421–432.

Rees, R. G., Thompson, J. P., and Mayer, R. J. (1979b). Slow rusting and tolerance to rusts in wheat. I. The progress and effects of epidemics of *Puccinia graminis tritici* in selected wheat cultivars. *Aust. J. Agric. Res.* **30,** 403–419.

Reiss, J. (1972). Cytochemischer Nachweis von Hydrolasen in Pilzzellen. II. Aminopeptidase. *Acta Histochem.* **39,** 277–285.

Roane, C. W., Stakman, E. C., Loegering, W. Q., Steward, D. M., and Watson, W. M. (1960). Survival of physiological races of *Puccinia graminis* var. *tritici* on wheat near barberry bushes. *Phytopathology* **50,** 40–44.

Röbbelen, G., and Sharp, E. L. (1978). "Mode of Inheritance, Interaction, and Application of Genes Conditioning Resistance to Yellow Rust." Verlag Paul Parey, Berlin and Hamburg.

Roberts, B. J., and Moore, M. B. (1956). The effects of temperature on the resistance of oat stem rust conditioned by the BC genes. *Phytopathology* **46,** 584.

Robinson, R. A. (1979). Permanent and impermanent resistance to crop parasites; a re-examination of the pathosystem concept with special reference to rice blast. *Z. Pflanzenzucht.* **83,** 1–39.

Robinson, R. A. (1980). New concepts in breeding for disease resistance. *Annu. Rev. Phytopathol.* **18,** 189–210.

Roelfs, A. P. (1974). Evidence for two populations of wheat stem and leaf rust in the U.S.A. *Plant Dis. Rep.* **58,** 806–809.

Roelfs, A. P., and McVey, D. V. (1974). Races of *Puccinia graminis* f. sp. *tritici* in the U.S.A. during 1973. *Plant Dis. Rep.* **58,** 608–611.

Roelfs, A. P., and McVey, D. V. (1975). Races of *Puccinia graminis* f. sp. *tritici* in the U.S.A. during 1974. *Plant Dis. Rep.* **59,** 681–685.

Roelfs, A. P., and McVey, D. V. (1976). Races of *Puccinia graminis* f. sp. *tritici* in the U.S.A. during 1975. *Plant Dis. Rep.* **60,** 656–660.

Roelfs, A. P., and Rothman, P. G. (1971). Races of *Puccinia graminis* f. sp. *avenae* in the U.S.A. during 1970. *Plant Dis. Rep.* **55,** 992–996.

Roelfs, A. P., and Rothman, P. G. (1972). Races of *Puccinia graminis* f. sp. *avenae* in the U.S.A. during 1971. *Plant Dis. Rep.* **56,** 608–611.

Roelfs, A. P., and Rothman, P. G. (1973). Races of *Puccinia graminis* f. sp. *avenae* in the U.S.A. during 1972. *Plant Dis. Rep.* **57,** 754–756.

Roelfs, A. P., and Rothman, P. G. (1974). Races of *Puccinia graminis* f. sp. *avenae* in the U.S.A. during 1973. *Plant Dis. Rep.* **58,** 605–607.

Roelfs, A. P., and Rothman, P. G. (1975). Races of *Puccinia graminis* f. sp. *avenae* in the U.S.A. during 1974. *Plant Dis. Rep. 59,* 614–616.

Roelfs, A. P., and Rothman, P. G. (1976). Races of *Puccinia graminis* f. sp. *avenae* in the U.S.A. during 1975. *Plant Dis. Rep.* **60,** 703–706.

Roelfs, A. P., Long, D. L., Casper, D. H., and McVey, D. V. (1977a). Races of *Puccinia graminis* f. sp. *tritici* in the U.S.A. during 1976. *Plant Dis. Rep.* **61,** 987–991.

Roelfs, A. P., Long, D. L., Casper, D. H., and Rothman, P. G. (1977b). Races of *Puccinia graminis* f. sp. *avenae* in the U.S.A. during 1976. *Plant Dis. Rep.* **61,** 1071–1073.

Roelfs, A. P., Casper, D. H., and Long, D. L. (1978a). Races of *Puccinia graminis* f. sp. *avenae* in the United States during 1977. *Plant Dis. Rep.* **62,** 600–604.

Roelfs, A. P., Casper, D. H., and Long, D. L. (1978b). Races of *Puccinia graminis* f. sp. *tritici* in the U.S.A. during 1977. *Plant Dis. Rep.* **62,** 735–739.

Roelfs, A. P., Casper, D. H., and Long, D. L. (1979a). Races of *Puccinia graminis* f. sp. *tritici* in the U.S.A. during 1978. *Plant Dis. Rep.* **63,** 701–704.

Roelfs, A. P., Casper, D. H., and Long, D. L. (1979b). Races of *Puccinia graminis* f. sp. *avenae* in the United States during 1978. *Plant Dis. Rep.* **63,** 748–751.

Rohringer, R., Howes, N. K., Kim, W. K., and Samborski, D. J. (1974). Evidence for a gene-specific RNA determining resistance in wheat to stem rust. *Nature* (*London*) **249,** 585–587.

Roseman, S. (1970). The synthesis of complex carbohydrates in multiglycosyl transferase systems and their potential functions in intercellular adhesion. *Chem. Phys. Lipids* **5,** 270–297.

Rosen, H. R. (1949). Oat percentage and procedures for combining resistance to crown rust, including race 45, and Helminthosporium blight. *Phytopathology* **39,** 20.

Rosen, H. R. (1955). New germ plasm for combined resistance to Helminthosporium blight and crown rust of oats. *Phytopathology* **45,** 219–221.

Rowell, J. B., and Roelfs, A. P. (1971). Evidence for an unrecognized source of overwintering wheat stem rust in the United States. *Plant Dis. Rep.* **55,** 990–992.

Russell, G. E., Andrews, C. R., and Bishop, C. D. (1976). Development of powdery mildew on leaves of barley varieties at different growth stages. *Ann. Appl. Biol.* **82,** 467–476.

Russell, W. A. (1961). A comparison of five types of testers in evaluating the relationship of stalk rot resistance in corn inbred lines and stalk strength of the lines in hybrid combinations. *Crop Sci.* **1,** 393–397.

Russell, W. A. (1965). Effect of corn leaf rust on grain yield and moisture in corn. *Crop Sci.* **5,** 95–96.

Samborski, D. J. (1972a). Leaf rust of wheat in Canada in 1971. *Can. Plant Dis. Surv.* **52,** 8–10.

Samborski, D. J. (1972b). Leaf rust of wheat in Canada in 1972. *Can. Plant Dis. Surv.* **52,** 168–170.

Samborski, D. J. (1974). Leaf rust of wheat in Canada in 1973. *Can. Plant Dis. Surv.* **54,** 8–10.

Samborski, D. J. (1975). Leaf rust of wheat in Canada in 1974. *Can. Plant Dis. Surv.* **55,** 58–60.

Samborski, D. J. (1976a). Leaf rust of wheat in Canada in 1975. *Can. Plant Dis. Surv.* **56,** 12–14.

Samborski, D. J. (1976b). Leaf rust of wheat in Canada in 1976. *Can. Plant Dis. Surv.* **56,** 123–124.

Samborski, D. J. (1978). Leaf rust of wheat in Canada in 1977. *Can. Plant Dis. Surv.* **58,** 53–54.

Samborski, D. J. (1979). Leaf rust of wheat in Canada in 1978. *Can. Plant Dis. Surv.* **59,** 67–68.

Samborski, D. J., and Dyck, P. L. (1976). Inheritance of virulence in *Puccinia recondita* on six backcross lines of wheat with single genes for resistance to leaf rust. *Can. J. Bot.* **54,** 1666–1671.

Samborski, D. J., Rohringer, R., and Kim, W. K. (1978). Transcription and translation in diseased plants. *In* "Plant Disease" (J. G. Horsfall and E. B. Cowling, eds.), Vol. III, pp. 375–390. Academic Press, New York.

Sanghi, A. K., and Luig, N. H. (1974). Resistance in three common wheat cultivars to *Puccinia graminis*. *Euphytica* **23,** 273–280.

Saxena, K. M. S., and Hooker, A. L. (1968). On the structure of a gene for resistance in maize. *Proc. Nat. Acad. Sci. U.S.A.* **61,** 1300–1305.

Scandalios, J. G. (1974). Isozymes in development and differentiation. *Annu. Rev. Plant Physiol.* **25,** 225–258.

Schnathorst, W. C., and De Vay, J. E. (1963). Common antigens in *Xanthomonas malvacearum* and *Gossypium hirsutum* and their possible relationships to host specificity and disease resistance. *Phytopathology* **53,** 1143.

Scott, P. R., Johnson, R., Wolfe, M. C., Lowe, H. J. B., and Bennett, F. G. A. (1979). Host-specificity in cereal parasites in relation to their control. *Plant Breeding Inst.* (*Cambridge*) Annu. Rep. for 1978, 27–62.

Sedcole, J. R. (1978). Selection pressures and plant pathogens: Stability of equilibria. *Phytopathology* **68,** 967–970.

Seevers, P. M., and Daly, J. M. (1970). Studies on the wheat stem rust resistance controlled by the *Sr*6 locus. *Phytopathology* **60,** 1642–1647.

Seevers, P. M., Daly, J. M., and Catedral, F. F. (1971). The role of peroxidase isozymes in resistance to wheat stem rust disease. *Plant Physiol.* **48,** 353–360.

Sequeira, L. (1978). Lectins and their role in host-pathogen specificity. *Annu. Rev. Phytopathol.* **16,** 453–481.

Shaner, G., and Hess, F. D. (1978). Equations for integrating components of slow leaf-rusting resistance in wheat. *Phytopathology* **68,** 1464–1469.

Shattock, R. C. (1976). Variation in *Phytophthora infestans* on potatoes grown in walk-in polyethylene tunnels. *Ann. Appl. Biol.* **82,** 227–232.

Shattock, R. C., Janssen, B. D., Whitbread, R., and Shaw, D. S. (1977). An interpretation of the frequencies of host specific phenotypes of *Phytophthora infestans* in North Wales. *Ann. Appl. Biol.* **86,** 249–260.

Shepherd, K. W., and Mayo, G. M. E. (1972). Genes conferring specific plant disease resistance. *Science* (*Washington, D.C.*) **175,** 375–380.

Shrum, R. D. (1978). Forecasting of epidemics. *In* "Plant Disease" (J. G. Horsfall and E. B. Cowling, eds.), Vol. II. pp. 223–238. Academic Press, New York.

Sidhu, G. S. (1975). Gene-for-gene relationships in plant parasitic systems. *Sci. Prog.* **62,** 467–485.

Simons, M. D. (1954). The relationship of temperature and stage of growth to the crown rust reaction of certain varieties of oats. *Phytopathology* **44,** 221–223.

Sinden, J. W. (1937). Bean anthracnose—a study in infection. Ph.D. Thesis, Cornell Univ. Ithaca, (Quoted by Mercer *et al.,* 1975).

Singleton, L. C., and Young, H. (1968). The *in vitro* culture of *Puccinia recondita* f. sp. *tritici*. *Phytopathology* **58,** 1068.

Smith, D. R. (1975). Expression of monogenic chlorotic-lesion resistance to *Helminthosporium maydis* in corn. *Phytopathology* **65,** 1160–1165.

Smith, D. R., and Hooker, A. L. (1973). Monogenic chlorotic-lesion resistance to *Helminthosporium maydis* in corn. *Crop Sci.* **13,** 330–331.

Stakman, E. C., and Harrar, J. G. (1957). "Principles of Plant Pathology." Ronald Press, New York.

Stakman, E. C., and Levine, M. N. (1922). The determination of biologic forms of *Puccinia graminis* on *Triticum* spp. *Minn. Agr. Exp. Sta. Tech. Bull.* **8,** 10 pp.

Stakman, E. C., Levine, M. N., Cotter, R. U., and Hines, L. (1934). Relation of barberry to the origin and persistence of physiologic forms of *Puccinia graminis*. *J. Agric. Res.* **48,** 953–969.

Stakman, E. C., Stewart, D. M., and Loegering, W. Q. (1962). Identification of physiologic races of *Puccinia graminis* var. *tritici*. *U.S. Dep. Agric. Agric. Res. Serv. Bull.* E617.

Statler, G. D. (1979). Inheritance of virulence of *Melampsora lini* race 218. *Phytopathology* **69,** 257–259.

Statler, G. D., and Zimmer, D. E. (1976). Inheritance of virulence of race 370 of *Melampsora lini*. *Can. J. Bot.* **54,** 73–75.

Stellwagen, E., and Wilgus, H. (1978). Relationship of protein thermostability to accessible surface area. *Nature* (*London*) **275,** 242–243.

Stevenson, F. T., Akeley, R. V., and Webb, R. E. (1955). Reactions of potato varieties to late blight and insect injury as reflected in yields and percentage solids. *Am. Potato J.* **32,** 215–221.

Strobel, G. A., and Sharp, E. L. (1965). Protein of wheat associated with infection type of *Puccinia striiformis*. *Phytopathology* **55,** 413–414.

Tanford, C. (1973). "The Hydrophobic Effect: Formation of Micelles and Biological Membranes." Wiley, New York.

Tanford, C. (1978). The hydrophobic effect and the organization of living matter. *Science* (*Washington, D.C.*) **200,** 1012–1018.

Tewari, J. P., and Skoropad, W. P. (1976). Relationship between epicuticular wax and blackspot caused by *Alternaria brassicae* in three lines of rapeseed. *Can. J. Plant Sci.* **56,** 781–785.

Thompson, J. N. (1975). Quantitative variation and gene number. *Nature* (*London*) **258,** 665–668.

Thresh, J. M. (1978). The epidemiology of plant virus diseases. *In* "Plant Disease Epidemiology" (P. R. Scott and A. Bainbridge, eds.), pp. 79–91. Blackwell, Oxford.

Thurston, H. D. (1971). Relationship of general resistance: late blight of potatoes. *Phytopathology* **61,** 620–626.

Turel, F. L. M. (1969). Saprophytic development of flax rust, *Melampsora lini,* race no. 3. *Can. J. Bot.* **47,** 821–823.

Turner, M. T., and Johnson, E. R. (1980). A race of *Helminthosporium turcicum* not controlled by *Ht* genetic resistance in corn in the American corn belt. *Plant Dis.* **64,** 216–217.

Ullstrup, A. J., and Brunson, A. M. (1947). Linkage relationships of a gene determining susceptibility to a Helminthosporium leaf spot. *J. Am. Soc. Agron.* **39,** 606–609.

Underwood, J. F., Kingsolver, C. H., Peet, C. E., and Bromfield, K. R. (1959). Epidemiology of stem rust of wheat. III. Measurements of increase and spread. *Plant Dis. Rep.* **43,** 1154–1159.

Uritani, I. (1971). Protein changes in diseased plants. *Annu. Rev. Phytopathol.* **9,** 211–234.

Vanderplank, J. E. (1963). "Plant Diseases: Epidemics and Control." Academic Press, New York.

Vanderplank, J. E. (1965). Dynamics of epidemics of plant disease *Science* (*Washington, D.C.*) **147,** 120–124.

Vanderplank, J. E. (1968). "Disease Resistance in Plants." Academic Press, New York.

Vanderplank, J. E. (1971). Stability of resistance to *Phytophthora infestans* in cultivars without *R* genes. *Potato Res.* **14,** 263–270.

Vanderplank, J. E. (1975). "Principles of Plant Infection." Academic Press, New York.

Vanderplank, J. E. (1978). "Genetic and Molecular Basis of Plant Pathogenesis." Springer-Verlag, Berlin and New York.

Vohl, G. T. (1938). Untersuchungen über den Braunrost des Weizens. *Z. Pflanzenzücht.* **22,** 233–270.

Von Broembsen, S. L., and Hadwiger, L. A. (1972). Characterization of disease resistance responses in certain gene-for-gene interactions between flax and *Melampsora lini*. *Physiol. Plant Path.* **2,** 207–215.

Walker, J. C. (1930). Inheritance of *Fusarium* resistance in cabbage. *J. Agric. Res.* **40,** 721–745.

Walker, J. C. (1966). The role of pest resistance in new varieties. *In* "Plant Breeding" (K. J. Frey, ed.), pp. 219–242. Iowa State Univ. Press, Ames.

Wallace, H. R. (1978). Dispersal in time and space: soil pathogens. *In* "Plant Disease" (J. G. Horsfall and E. B. Cowling, eds.), Vol. II, pp. 181–202. Academic Press, New York.

Wallin, J. (1964). Texas, Oklahoma and Kansas winter temperatures and rainfall, and summer occurrence of *Puccinia graminis tritici* in Kansas, Dakotas, Nebraska and Minnesota. *Int. J. Biometeorol.* **7,** 241–244.

Warren, R. C., King, J. E., and Colhoun, J. (1971). Reaction of potato leaves to infection by *Phytophthora infestans* in relation to position on the plant. *Trans. Br. Mycol. Soc.* **57,** 501–514.

Warren, R. C., King, J. E., and Colhoun, J. (1973). Reaction of potato plants to *Phytophthora infestans* in relation to their carbohydrate content. *Trans. Br. Mycol. Soc.* **61,** 95–105.

Waterhouse, W. L. (1929). Australian rust studies. *Proc. Linnean Soc. N.S. Wales* **5,** 615–680.

Watson, I. A. (1958). The present status of breeding disease resistant wheats in Australia. Farrer Oration. *Agric. Gaz. N.S. Wales* **59,** 630–660.

Watson, I. A., and Luig, N. H. (1963). The classification of *Puccinia graminis* var. *tritici* in relation to breeding resistant varieties. *Proc. Linnean Soc. N.S. Wales* **88,** 235–258.

Watson, I. A., and Luig, N. H. (1966). *Sr*15—A new gene for use in the classification of *Puccinia graminis* var. *tritici*. *Euphytica* **15,** 239–250.

Watson, I. A., and Luig, N. H. (1968). Progressive increase of virulence in *Puccinia graminis* f. sp. *tritici*. *Phytopathology* **58,** 70–73.

Webb, R. E., and Bonde, R. (1956). Physiological races of the late blight fungus from potato dump-heap plants in Maine in 1955. *Am. Potato J.* **33,** 53–55.

Williams, P. G., Scott, K. J., and Kuhl, J. L. (1966): Vegetative growth of *Puccinia graminis* f. sp. *Tritici in vitro*. *Phytopathology* **56,** 1418–1419.

Wolfe, M. S. (1978). Some practical implications of the use of cereal variety mixtures. *In* "Plant Disease Epidemiology" (P. R. Scott and A. Bainbridge, eds.), pp. 201–207. Blackwell, Oxford.

Zacharius, R. M., and Osman, S. F., Heisler, E. G., and Kissinger, J. C. (1976). Effect of the R-3 gene in resistance of the Wauseon potato tuber to *Phytophthora infestans*. *Phytopathology* **66,** 964–966.

Zimmer, D. E., and Schafer, J. F. (1961). Relation of temperature to reaction type of *Puccinia coronata* on certain oat varieties. *Phytopathology* **51,** 202–203.

Index

O

P

Q

R

X